CRUSHED

BIG TECH'S WAR ON FREE SPEECH

KEN BUCK

Humanix Books

www.humanixbooks.com

To Hannah, Bear, Sugar Ray, Dubya, T-Man, Collins, and Doe.

As young Bucks, you exercised your free speech rights anytime you wanted. Your strength of personality to overcome efforts that would suppress the content or volume of your strongly held convictions has inspired me to write this book so that Big Tech can't accomplish what your parents failed to do.

Contents

Foreword

Big Tech poses the single greatest threat to free speech and democracy in our country. My friend Ken Buck has assembled a must-read dossier that lays out the enormous dangers we face as a nation right now.

Crushed exposes Big Tech's most egregious abuses of power—from censorship, discrimination, and political favoritism to self-preferencing, self-dealing and predatory, anticompetitive efforts. Any other industry that trampled free speech rights, destroyed the playing field for competition, and silenced politicians and charity groups for their political views would face a mountain of lawsuits for their brazen abuses of power.

These companies have spent tens of millions of dollars on campaigns to stop Congress from leveling the playing field for all, and they have entrenched Silicon Valley lackeys in key government regulatory positions.

Newspaper magnate William Randolph Hearst would be stunned by Big Tech's near-total control over communication and the dissemination of information. John D. Rockefeller would marvel at Big Tech's economic control of the marketplace. Big Tech companies are the robber barons of the twenty-first century. As the adage goes, "With great power comes great responsibility." But what has Big Tech done? Abused that power at every turn.

There has never been such an aggregation of power in the history of humankind as Big Tech enjoys today with the money, monopoly power, and hubris that comes with the unchecked exercise of power.

This book arrives resounding like a five-alarm warning and explains the threats we face in great detail.

Ken demonstrates his commitment to open discourse, he explains how monopolies actively censor and restrict free markets, and he shows how society suffers from anticompetitive tactics. In *Crushed*, Ken details the predatory, bullying, and biased behavior of Amazon, Apple, Facebook, and Google.

A dedicated conservative, Ken is a tireless champion for freedom and liberty. I am grateful for his work on behalf of the American people. His book is a public service, and he does a masterful job of connecting the dots. Ken makes the unassailable case that we must unite to stop Big Tech billionaires from censoring speech, stopping the exchange of ideas, manipulating voters, filtering information, and quashing their political opponents.

U.S. Senator Ted Cruz, R-Texas

CHAPTER 1

Censoring Political Debate

CONSERVATIVE IDEAS UNDER ATTACK

If all mankind minus one were of one opinion,
mankind would be no more justified in silencing
that one person than he, if he had the power,
would be justified in silencing mankind.

JOHN STUART MILL

On the evening of Wednesday, September 22, 2010, I flew into Dallas/Fort Worth Airport. I had three fundraisers to attend the following day. Breakfast in Dallas, lunch in Austin, and dinner in Houston. With all that meeting, greeting, and eating in the near future, I made it a point to get some exercise. I checked into my hotel, changed, and headed to the gym for a quick workout.

I got on the treadmill. It was one of those high-tech, foo-foo machines with a TV screen in the middle of the control panel. I didn't have any headphones, so I just started the machine to do what it was built for—working out.

There was a younger guy on the machine next to me. He had planned ahead and brought earphones, so he was locked into the screen in front of him. I didn't mind. I wanted to run, not chat.

Running for election is work. Running for myself is not.

Then I glanced to my left and got a shock: my face was on his TV.

I looked a little harder and saw a caption below my face. It said: "World's Worst American."

Of course, I couldn't hear what was being said about me. But I knew it wasn't anything good with a caption like that.

So, I just kept running.

My neighbor took a glance at me that turned into a double take. I could see him looking back and forth as he connected me to the "World's Worst American" on his mini-TV.

Then my image was replaced by liberal commentator Keith Olbermann. For those too young to remember, Olbermann was

3

MSNBC's left-wing pit bull. He was nasty. He was mean-spirited. He made a career passing judgment on everyone but himself. My neighbor hit the stop button on his treadmill and continued watching the show as his machine slowed. He ripped his headphone from the machine while shaking his head. Finally, he looked in my direction and made eye contact, rewarding me with a nasty sneer before leaving the gym.

I continued running. I didn't find out until much later what the segment was about, but the irritated jogger hadn't asked me if any of what Olbermann had said was true. He just assumed it was and stomped off.

But that's democracy. The only place where people can't make unfair, partisan, and vicious accusations against the people they disagree with politically are in authoritarian regimes. In America, Olbermann can say what he likes, and the treadmill guy can believe what he likes. This is what makes America great.

———

I don't mind what happened in Dallas or in MSNBC's New York studio. I think it's worth celebrating—and not just because I view getting insulted by a knee-jerk liberal on MSNBC as a badge of honor.[1] No, while I always advocate for dialogue, civility, analysis, and understanding over spite and condemnation, I'm okay with people who don't like what I say.

That's called supporting free speech. That's called being an American.

I'm happy to be insulted on a reality-challenged liberal TV network because I believe in the right to the uncensored exchange of ideas and information. After all, that's the essential principle upon which this country was founded.

Olbermann has every right to say what he believes—as long as it isn't libelous, obscene, or inciting violence. And I have the right to do the same. Meanwhile, our fellow citizens have the right to judge the ideas we espouse and make their own determinations.

Unfortunately, free speech and the exchange of ideas—frequently, *conservative ideas*—are under attack in America.

These assaults, often perpetuated by all-powerful Big Tech companies and liberal activists who should know better, are indirect attacks on the two things that lie at the core of America's strength: our democracy and our economy. We need to fix this to protect our present and to ensure our future. Or as the preamble of the Constitution puts it, to "secure the Blessings of Liberty to ourselves and our Posterity."

THE MARKETPLACE OF IDEAS

In 1859, political economist John Stuart Mill wrote a book called *On Liberty*. Chapter 2 focused on the importance of the free flow of ideas. Mill believed that ideas should compete against one another for the good of all. In his view, every idea—from theories and policy to inventions and products—should be evaluated by society to determine their "truth" or "effectiveness."[2]

For Mill, ensuring an unimpeded flow was a way to protect individual independence and prevent social control by a government or an oppressive popular idea. Supreme Court Justice Oliver Wendell Holmes built on Mill's ideas while writing a dissenting opinion in 1919, "The ultimate good desired is better reached by free trade in ideas—that the best test of truth is the power of the thought to get itself accepted in the competition of the market."[3]

Over time, Mill's and Holmes's views have become encapsulated by a single, vital phrase: the marketplace of ideas. This marketplace has no physical location, of course. It is an abstract but very real term for something that exists or should exist in free, open societies—where ideas gain acceptance by competing with one another, without the threat of censorship, to prove their value.

Competition in the marketplace of ideas is based on quality, on evaluation, on free and unfettered exchange with whatever communication methods are available. Recently, however, as Big Tech companies consolidate power and influence our leaders, America is now a place where too often opinions are stifled or labeled "dangerous." Public and private figures are silenced and even banned from digital platforms. Giant corporations use their market share power and technology to squash competitors, stealing rival technology, or simply buying an innovative rival and shutting it down.

This behavior stops the flow of ideas. It shrinks and even shutters both the marketplace of ideas and the marketplace of commerce. In doing so, it destroys competition, which, in turn, destroys our social and economic growth.

A STORM OF QUIETING

Until recently most Americans probably didn't think censorship occurred within our borders. They thought it happened in other places—China, Saudi Arabia, Russia, North Korea, Iran, Myanmar. These repressive countries limit and monitor speech to frightening degrees. The Chinese Communist Party has thwarted any talk of democratic reform for decades, most recently crushing protest and dissent in Hong Kong. It has created a giant firewall to prevent its citizens from reading foreign websites criticizing its leaders or policies. As I was finishing this book, Russian leader Vladimir Putin signed a law that would punish anyone publishing "fake news" with up to 15 years in jail—a move with a chilling effect on Russian citizens and foreign journalists reporting in Russia.

Communist governments and centralized monarchies like Saudi Arabia are threatened by dissent. By new ideas. With power and wealth consolidated at the top, most citizens of these countries cannot rise socially and economically. These systems, which artificially control financial markets, cannot allow a marketplace of ideas to flourish because that marketplace is too threatening; it challenges governmental authority. The same goes for freedom of speech. The opposite of freedom is control or enslavement. This is why authoritarian and totalitarian, nondemocratic governments rely on censorship. Freedom is a threat to power.

I'm not grandstanding here. The American Dream—a phrase the whole world understands—is a dream of freedom, of fair play and financial potential. Is there a similar expression for authoritarian countries? I don't recall ever hearing anyone pining for the Russian, Chinese, or Saudi Dream.

When it comes to turning a dream into reality, the cornerstone of American growth and prosperity is the First Amendment. It's worth emphasizing these guarantees were listed first for a reason: our founders believed that guaranteeing freedom of religion, speech, the press, assembly, and the right to petition for change was hugely important. You could say they provide a secure, positive foundation for the marketplace to thrive.

The American Dream is great, it's exceptional, but unfortunately, it's in deep peril. The American Dream is threatened—and specifically by Big Tech. Americans rightfully fear that we are on a path to become like Russia, China, or Saudi Arabia if we do nothing.

SILENCING IN PLAIN SIGHT

Ironically, this subtle censorship of speech and restrictions on the marketplace of ideas is spearheaded by four companies that most people think of as *promoting* speech. These Big Tech players rank among the most highly valued corporations in the world: Apple, Google, Facebook, and Amazon.

There is a fifth major player that belongs in this group: Twitter, a giant messaging, marketing, and amplification platform. Although it doesn't have the profits or valuation of its Big Tech buddies, Twitter is part of an ideologically driven cartel that claims to be open to all but selectively flexes its muscles and acts as a type of thought police.

On the surface, these companies facilitate communication. Apple and Google own the world's two dominant cellphone platforms: iOS and Android. Social media behemoth Meta, which owns Facebook, Instagram, and WhatsApp, is, along with Google, the world's biggest disseminator of information—news, ideas, opinions, images. It facilitates sharing of information (posts, photos, news articles, music, videos) to gather vast troves of personal data about their users and then sell that information to advertisers. Amazon dominates American retail, capturing two-thirds of online shopping traffic in the United States, and provides the backbone of the web for millions of businesses with Amazon Web Services, used for hosting websites and online storage.

I'm describing this Big Tech foursome in very broad strokes here—I'll zero in on specific abuses later—but the takeaway should be clear. Through market share, technology, and policy, Apple, Facebook, Google, and Amazon have acquired control over the essential infrastructure of America's commerce and communications. They are monopolies. Their concentrated power and foundational technologies make them the gatekeepers to the marketplace of ideas. They make the rules about what society sees: what information is added to a news feed, what apps are sold on their phones, what products are listed in their search results.

In other words, their power and their business models result in the selective dissemination of information and infringe on the free flow of ideas.

The fact that these companies have consolidated this vast amount of power should concern all citizens, regardless of party. But we know that these companies have exerted that power—both to benefit themselves economically and to suppress their ideological opponents in the marketplace of ideas. No one—conservative or liberal—should be comfortable with a few Silicon Valley oligarchs having a monopoly over the marketplace of ideas, and with it, democracy itself.

OLD STORY, NEW DANGERS

America has overcome enormous monopolies before. The wealthy robber barons of the late nineteenth and early twentieth centuries controlled energy, finance, steel, and commercial transportation. Those captains of commerce—John D. Rockefeller, J. P. Morgan, Andrew Carnegie, and Cornelius Vanderbilt—didn't control the flow of information, so they could only influence our democracy through bribes, political campaign corruption, and other political peddling schemes to ensure market control and profits. Yes, they also created foundations and charities that have rehabilitated their names. And in the same vein, it is important to bear in mind the Big Four have done immense good, creating platforms and services that have fueled enormous growth and often life-changing progress.

But that doesn't mean they haven't engaged in nefarious conduct. Far from it. Some economists believe we have now entered a second Gilded Age, with Big Tech as the new robber barons. They make a compelling case.

Like the robber barons of yesteryear, Big Tech monopolies threaten the core of our economic system by engaging in predatory pricing, exclusionary fees, anticompetitive takeovers, and more.

In contrast to the robber barons' incidental political corruption, Big Tech threatens the core of our political system by controlling what information is distributed to the public and disseminating or impeding that information to benefit their own commercial interests and political views. The huge troves of personal data these monopolies gather influence that distribution—to spur more engagement, sales, and often ideological reinforcement, all of which is designed to create a compelling, convenient, *seemingly harmless* user experience that keeps unwitting consumers coming back for more. Indeed, most users regard the Big Four and Twitter as platforms that provide useful, positive services, not censorious, digital gatekeepers.

They are, unfortunately, both.

POWER OVER FREEDOM

Big Tech is aware of its power and its innovative competitors. Hyper aware.

I have joined conservative leaders in the U.S. House and Senate to sponsor several bipartisan bills that are circulating through Congress focused on safeguarding the marketplace of ideas. In the upcoming chapters, we will examine specific anticompetitive conduct by Big Tech and the legislative solutions that would promote competition, enhance innovation, and protect free speech.

Big Tech companies are not sitting idly on the sidelines as passive observers. They are heavily invested in maintaining the status quo of monopoly control over their commercial interests and ideological goals.

They are fully engaged in political patronage—hiring the family members of elected leaders; making targeted political donations to the campaigns of critical members of Congress; buying off various Washington, D.C., think tanks, academic centers, and advocacy groups. Meta, the company that owns Facebook, spent $20 million on lobbying in 2021. Amazon spent more than $19 million.[4] The only public company that spent more on lobbying was Blue Cross/Blue Shield.

Why spend this money and make these hires? It's a gigantic effort to create a feedback loop of influence in Washington to stop any challenges to their massive market power.

That power—while it benefits one side of the political spectrum today—should concern all Americans. Big Tech remains unchecked, its power threatens the foundational ideas of America. For those who might think this is alarmist, let me be clear:

- When control over information in a democracy rests in the hands of only a few individuals, the results of an election can be manipulated by those individuals.
- When a few companies command and control critical digital media, they are positioned to dominate and distort the marketplace of ideas.
- When a new competitive product or technology threatens to disrupt the dominant position of an entrenched monopoly, that newcomer should be allowed to flourish—not be forced out of business or bought and shelved.
- When a company amasses so much revenue it can sell goods and services below cost to eliminate competition and dominate markets, it must be monitored to avoid predatory actions that, in the long run, are poised to stifle competition.

Before I got to Congress, all I remembered from my antitrust class in law school was that soon after the Civil War, Congress developed a legal structure to protect consumers from predatory conduct and promote competition between businesses. Congress did not micromanage this area of law and left the specific interpretation of these "antitrust laws" to the courts. As commerce changed, it became more difficult for the courts to adapt their judicial precedent to novel business practices. Every so often, the concentration of power within an industry became so lopsided that Congress stepped in to protect consumers and create openings for small competitors to challenge the entrenched monopolies.

My attitude toward the application of antitrust laws to Big Tech changed when I attended a hearing in January 2020, in my role as a member of the Antitrust Subcommittee of the House Judiciary Committee. I believe in free, unregulated markets. When I began to examine the evils perpetrated by Big Tech, I thought the market would correct itself.

After attending the hearing, studying the history of competition policy in America, and talking to several of my colleagues on both sides of the aisle, I now understand that Big Tech is so powerful it can and does control many aspects of the commercial marketplace and the marketplace of ideas in our society in a way that defies oversight by the market, courts, or government regulators. Dissenting voices have been thrown off Twitter and Facebook. Antiestablishment views about COVID-19 have been silenced. Posts have been censored. News articles have been buried in search results. Proprietorial data and intellectual property have been brazenly stolen in broad daylight and utilized by Big Tech. The marketplace of ideas is now a gated community within the digital sphere. For 200 years, that marketplace was, by and large, self-regulated. Not anymore.

Rather than regulate these monopolies, we need laws that promote competition, giving the free market a chance to actually flourish. We need to make sure monopolies do not weaponize their market dominance to stifle innovation and competition. Allowing such predatory behavior may harm our society, reducing consumer choices and raising prices. Some might suggest this is a form of big government regulation. However, that is a case of missing the forest for the trees. These monopolies already regulate what is said and what is seen, read, and digested without any input whatsoever from the American people. They determine what is sold and promoted on their platforms. Their algorithms determine what we see and what we don't see because they place relevance values on content. If an article, book, video, or photograph is given a low relevance rank, you may never learn of its existence. That kind of control can be leveraged to curry favor with or actively harm a person, politician, or party.

Influence in Action

During the run-up to the 2020 presidential election, both Facebook and Twitter actively prevented potentially damaging *New York Post* reports about Joe Biden's son Hunter from reaching the public. Reacting to reports about the contents of Hunter Biden's discarded laptop, including emails that show he introduced his father to a Ukrainian energy executive (something that the elder Biden had denied), Facebook representative Andy Stone said, "We are reducing [the *Post*'s] distribution on our platform." Twitter was even more censorious: it blocked users from posting links to the *Post*'s story and shut down the paper's Twitter account for two weeks.

Interestingly, a year later—after the election was over and Biden was in the Oval Office—Twitter CEO Jack Dorsey admitted the so-called hacking offense was nonexistent and silencing the *Post* was a "mistake."[5]

This was a clear case of Big Tech suppressing information that might have changed the outcome of the 2020 election. They were gatekeepers, selectively determining which stories made it into the marketplace of the news.

Similarly, Google's search engine is an incredible product. But its proprietary algorithmic relevance logic is, by definition, intrinsically exclusionary. Some results appear prominently at the top of the page, others are buried far below. While the powers that be at Google insist there is no political bias in its search engine, one 2018 report found that a news search on the term "Trump" returned an overwhelming number of articles from left-of-center outlets. The first page included two links to CNN, CBS, the *Atlantic*, CNBC, the *New Yorker*, and Politico. There were no right-leaning sites listed.[6]

The number of instances when Big Tech uses its monopoly to suppress opinions keeps growing, frequently targeting conservative politicians. Senator Rand Paul of Kentucky has found himself locked out of his Facebook page for "repeatedly going against our community standards," suspended from Twitter for questioning the effectiveness of masks, and had videos removed from YouTube, which is owned by Google. The year 2021 also saw a host of other suspensions. His colleague, Senator Ron Johnson from Wisconsin, was temporarily blocked from YouTube for comments about COVID treatments. And U.S. Representative Jim Banks of Indiana was suspended for two weeks after mocking *Time* magazine for giving the Woman of the Year award to a "biological male."[7]

Meanwhile, the former president of the United States, Donald Trump, has been suspended from Facebook, Instagram, YouTube, and Twitter. The man who received nearly 47 percent of the vote—over 74 million ballots—has been suspended from America's most popular internet platforms. Twitter decided his presence was a "risk of further incitement of violence" following the January 6 riots at the U.S. Capitol.

Ironically, Trump is a victim of Big Tech discovering and honing its own power. Social media platforms were credited with substantially assisting President Barak Obama's victory in the 2008 presidential election. Silicon Valley celebrated its ability to elect a young liberal upstart and defeat the entrenched Democratic Clinton machine. The lessons of 2008 were not lost on Donald Trump. He used the same social media platforms in 2016 to bypass the mainstream media and speak directly to the people. Trump's victory shook Big Tech companies. They realized that they had created a pathway for conservatives to bypass the liberal media organizations that traditionally blistered conservative candidates. Now that pathway is being selectively monitored.

SPEAKING OUT FOR DEMOCRACY

For 25 years, I was a federal prosecutor and district attorney. I was driven by a passion for justice—for trying to do the right thing. Now that I serve in Congress, that passion has only increased. As a free-market advocate, I have no problem when businesses seek to monetize human interaction. As an American, however, I have a huge problem when those businesses divide and censor my fellow citizens and seek to control the marketplace of ideas for their own profit.

For John Stuart Mill's marketplace of ideas to flourish, America needs free technology markets, a level playing field for innovators, and choices for information consumers. Big Tech impedes the free flow of information when it uses its monopoly power to crush competition, steal technology, and suppress innovation. The marketplace of ideas and the commercial marketplace are damaged, and America is ultimately the loser.

This book is for all Americans who believe in the democratic ideals of free speech and free markets and adhere to our nation-defining concept "that all men are created equal" and deserve equal opportunity and a level playing field. Thomas Jefferson, who famously created that declaration of equality, was not a big believer in big government. "Were we directed from Washington when to sow and when to reap," he wrote in his autobiography, "we should soon want bread." But Jefferson was also wary of unfair competition and pushed against any impulse to give Congress the power to allow monopolies. Patents and copyrights are as close as the government has ever come to empowering exclusive commercial endeavors. That's because our founders realized that patents and copyrights inspire creative innovation, as we will see in Chapter 4, but monopolies crush innovation and the free market.

Freedom of speech is an issue that concerns all Americans. The current and looming crisis posed by Big Tech should bridge any conservative–liberal divide. Giving the power to suppress or control speech to an unaccountable Silicon Valley "content moderator" should concern everyone: civil rights advocates, people of all faiths, Republicans and Democrats, Tea Party members and socialists—everyone on the political spectrum.

With Big Tech's massive financial resources and command of critical digital media, these companies are positioned to dominate and distort the marketplace of ideas. This threat to free speech is a risk that America can't afford. In the upcoming chapters, I want to delve into how we got to this precarious point. To do that, I will first examine the history of free speech, antimonopoly law, and how competition prevents collusion of big government and big business. I'll also detail the endless stream of outrageous abuses perpetrated by the Big Four.

Learning about these abuses is disturbing and enraging. But only by exploring the misdeeds and the vulnerabilities of our democracy can we take steps to protect it. Unlike the current Big Tech philosophy, my goal in these pages is not to divide us, but to point our nation toward a more stable, thriving future that is richer and fairer.

For all.

CHAPTER 2

America's Heritage

FREE MARKETS AND FREE SPEECH

———————————

Congress shall make no law respecting an establishment
of religion, or prohibiting the free exercise thereof;
or abridging the freedom of speech, or of the press;
or the right of the people peaceably to assemble, and to
petition the Government for a redress of grievances.

FIRST AMENDMENT

As kids we were taught that the American Revolution was fought because the colonists suffered from taxation without representation. While true, the reality was much more complex and relevant to where we find ourselves today; monopolies by their nature restrict the business marketplace and the marketplace of ideas. Before there was Amazon, there was the East India Company, the corporation with a giant trading monopoly that extended throughout the British Empire and which, inadvertently, detonated the American Revolution.

Granted a royal charter on the last day of 1600 by Queen Elizabeth I, the East India Company began that new century with a clear mission: break the Spanish stranglehold on the spice trade and make piles of money. The charter essentially gave a group of well-heeled merchants trade rights to and "from all the Islands, Ports, Havens, Cities, Creeks, Towns or Places of Asia, Africa or America" to England and vice versa. It also, not incidentally, gave them firm war powers, if needed. The founding merchants' 70,000-pound investment soon evolved into the most powerful company on the planet.

Just as Amazon launched as a bookseller, the East India Company got its start primarily focusing on spices. By 1613, the company cut a deal with Mughal emperor Jahangir, giving it access to vast swaths of India. But why buy and sell only one product when you can sell everything everywhere? As the company expanded farther east to China and west to America and the West Indies, it participated firsthand in the evil of trafficking slaves. In addition to slaving, the company focused on textiles such as silk and the aromatic leaves of an evergreen shrub: the tea plant.

Under its charter, the East India Company sold tea as a wholesaler: it imported shipments to England and auctioned them to other merchants. Those merchants then sold the tea in England and across the empire, exporting it to places like the American colonies, where East India Company–wrangled tea was the only tea legally for sale.

As commodities go, tea in colonial America was a trendy item. It was more than just a tasty, warming, caffeinated beverage. It was a status symbol and it wasn't cheap,

The Townshend Acts, passed in British Parliament in 1767, assessed an import tax on the little leaves, as well as lead, glass, paper, and paint. That import duty was funneled straight back to the London treasury. Increasing the price of tea created a market for tea smuggled into the colonies by Dutch traders.

The competition from smugglers posed a considerable problem for what had been, up to that point, the most profitable commercial business enterprise—and monopoly—in history.[1] By 1770, Dutch competition and overaggressive purchases left the company with a severe inventory problem: it had too much tea—17 million pounds of the stuff—stored in its warehouses in England and not enough buyers. Nobody used the phrase back then, but a monopoly the size of the East India Company was deemed too big to fail by government bigwigs. Lord North, England's prime minister, ushered through the Tea Act, allowing the behemoth to ship its tea directly to the colonies, which lowered the price by cutting out the middleman. Additionally, the act lowered the tax rate for East India Company tea, while rival tea imports were taxed at a higher rate. Meanwhile, the Townshend Act's tea levy remained in place. This was crony capitalism in action: the monopoly could now sell more tea directly and at a lower cost to the colonies while the British government that facilitated those direct sales still reaped tax revenue.

You might think that colonists appreciated this monopoly-loving sweetheart deal because it resulted in lower tea prices. That perspective would be completely understandable given that under the current definition of the so-called consumer welfare standard, many courts in this country would find that the East India Company was acting in the best interests of the consumer and should be found blameless. Robert Bork first wrote about the consumer welfare standard in 1978 as a way to determine whether business practices violated antitrust law. Over the years, the courts have narrowed its original intent—to determine full economic efficiency—and often limit the analysis to whether consumer prices rise in the short term. Because Big Tech offers "free services," regulators and the courts have often failed to enforce basic antitrust law. One competitor to Big Tech has gone so far as to suggest that the consumer welfare standard has become a corporate welfare standard allowing Big Tech to insulate themselves from competition.[2]

Thankfully the consumer welfare standard didn't exist in the 1770s. The colonists used common sense to figure out what today's courts are unable to understand, that monopoly control in a market can result in various injuries beyond merely the short-term price of a product.

Many colonists worried about the Tea Act because they realized that there might be no limit to the power of the East India monopoly. Three years earlier, the company flexed its monopoly powers in drought-stricken Bengal, refusing to ease tariffs while hoarding food and supplies for its troops. Over a million people perished from the severe crisis, but the company thrived.

The company's callous behavior was not lost on the colonists, least of all on John Dickinson, a lawyer who split his time between homes in Philadelphia and Delaware. On November 27, 1773, writing under the name Rusticus, Dickinson published an open letter printed as a broadside, a large sheet of paper, lambasting the new tea tax and listing the East India Company's sins in black and white. Two hundred fifty years later, it remains a powerful list of grievances:

> Their conduct in Asia, for some years past, has given ample proof, how little they regard the law of nations, the rights, liberties, or lives of men. They have levied war, excited rebellions, dethroned princes, and sacrificed millions for the sake of gain. The revenue of mighty kingdoms have centered in their coffers. And these not being sufficient to glut their avarice, they have, by the most unparalleled barbarities, extortions and monopolies, stripped the miserable inhabitants of their property and reduced whole provinces to indigence and ruin. Fifteen hundred thousand perished by famine in one year, not because the earth denied its fruits, but this company and its servants engrossed all the necessities of life, and set them at so high a rate, that the poor could not purchase them. Thus having drained the sources of that immense wealth, they now, it seems, cast their eyes to America, as a new threat, whereupon to exercise their talents of rapine, oppression and cruelty. The monopoly of tea is, I dare say, but a small part of the plan they have formed to strip us of our property.[3]

Dickinson's concerns resonated and reflected the fears of many colonists. If one company could control the flow of a specific product, what was stopping it from broadening control of other goods and markets? On December 16, just three weeks after Dickinson published this critique, angry colonists led by the Sons of Liberty, a group that counted John Hancock and Samuel Adams as members, went to Griffin's Wharf in Boston and dumped 342 chests of tea into the harbor.

This, of course, was the Boston Tea Party. Most Americans learn about this event at an early age. The key chorus logged in our collective memory—"no taxation without representation"—is widely considered the crux of the protest. That was doubtlessly part of the story, but it misses the complete picture. As many scholars now note, the Tea Party in Boston was motivated more by small businesses furious at the power of a giant, competition-quashing global monopoly than the larger issue of being second-class citizens with no advocates in Parliament.

It's also important to remember the anger at the monopoly had spread far and wide. Dickinson, sometimes called "the pen-man of the Revolution," was in Philadelphia, and a few months later, Sons of Liberty in New York staged its own tea party. The rebellion, then, was a colony-wide movement driven by fear of monopoly power and amplified through the power of speech and the press.

THE PRO-MONOPOLY BRITISH ARE COMING

The act of dumping tea was a thunderous, high-volume statement with a message that local businesses would not tolerate anticompetitive, monopolistic laws. The British Parliament's response to that statement shows just how connected monopolies are to censorship—or how freedom of speech is rooted in free markets.

After the Tea Party, the politicos of Westminster enacted the Coercive Acts in 1774, a quartet of laws punishing Massachusetts. From across the Atlantic, Parliament shut down the port of Boston, restricted democratic town meetings and turned the governor's council into an appointed body, made British officials immune to criminal prosecution in Massachusetts, and required colonists to house and quarter British troops on demand. This type of disenfranchisement is part and parcel of authoritarian behavior. When government officials are above the law, they are the law. When your home can be seized by the military, it ceases to be your home. When the elite receive preferential treatment under the law, there is no rule of law.

These restrictive measures set a direct course for the foundation of America. It took two more years of authoritarian abuse for a righteous and rebellious new nation to crystallize. But when it did arrive, it was announced with a stunning bang of rhetorical eloquence, economy, and insight. The Declaration of Independence opens with a glorious salvo stating the immediate presence of "self-evident" truths: *that all men are created equal, that they are endowed by their Creator with certain unalienable Rights, that among these are Life, Liberty, and the pursuit of Happiness.*

The Declaration also catalogs the examples of tyranny that the colonists could no longer abide, "a long train of abuses and usurpations." Although the word "monopoly" doesn't appear in the Declaration, its shadow looms behind the many charges leveled at the king, from the damning opening indictment, "He has refused his Assent to Laws, the most wholesome and necessary for the public good," to these side-by-side blasts: "For cutting off our Trade with all parts of the world" and "For imposing Taxes on us without our Consent"—both of which echo the disdain for favors given to the East India Company. The 27 instances cited by Jefferson illustrated the result of too much power in the monarch's hands.

It's easy to think of the Declaration of Independence as only a written document. But it's not just a piece of parchment. Even more than the Tea Party actions, it is a statement meant to be spoken: a throw-down about freedom and self-governance.

And, of course, to the ruling British government, the Declaration of Independence was nothing short of treason. It was a declaration of war.

Jefferson and Madison: On Liberty and Monopoly

Ten years after announcing America's independence, in a December 20, 1787, letter to James Madison, Thomas Jefferson was still thinking about freedom and monopolies and protecting citizens against government repression. He praised and then critiqued Madison's work on the Constitution:

> I will now add what I do not like. First the omission of
> a bill of rights providing clearly and without the aid of
> sophisms for freedom of religion, freedom of the press,
> protection against standing armies, restriction against
> monopolies, the eternal and unremitting force of the
> habeas corpus laws, and trials by jury in all matters of
> fact triable by the laws of the land. . . .[4]

Later in the letter he states: "[A] bill of rights is what the people are entitled to against every government on earth, general or particular, and what no just government should refuse." Jefferson's words hit a nerve in the marketplace of ideas that circulated in the early Congress and gained traction with federalists and anti-federalists alike.

James Madison responded to Jefferson: "Monopolies are sacrifices of the many to the few. Where the power is in the few it is natural for them to sacrifice the many to their own partialities and corruptions." And then he got down to business, writing the Bill of Rights. Liberty and equality—that vital template of self-evident terms that have, albeit often far too slowly, guided the evolution of American society—were front and center in his draft. As I noted earlier, the First Amendment protects speech, the press, and religion, basically establishing that the marketplace of ideas must remain open and devoted to all subjects.

Despite Jefferson's criticism of monopolies, Madison believed that the protection from monopolies would naturally flow from the new form of government that they were creating and therefore, unnecessary to include in the Constitution. He responded to Jefferson that "representational government at the federal level would prevent a repeat of the English experience with monopolies"[5] and that individual rights intrinsically protected against them. A just government, he wrote, is not "property secure . . . where arbitrary restrictions, exemptions, and monopolies deny to part of its citizens that free use of their faculties, and free choice of their occupations."[6]

While the word "monopoly" does not appear in the Constitution, antimonopoly sentiment is clear in the text. In Article IV, Section 2, Madison references the concept of privileges, which was broadly understood at the time to prohibit monopolies. This sentence—"The Citizens of each State shall be entitled to all Privileges and Immunities of Citizens in the several States"—is the key. "Privileges," under English law, included the right to be free of monopolies.[7] Freemen had the right to participate in mercantile endeavors on equal legal grounds with others. Monopolies, then, violated that right.[8]

The authors of the Articles of Confederation borrowed this idea, writing, "The Inhabitants of each Colony shall henceforth always have the same Rights, Liberties, Privileges, Immunities and Advantages in the other Colonies, which the said Inhabitants now have." In other words, they regarded the absence of monopolies as settled law.

THE PRINTING PRESS AND THE INTERNET

The big tech of the colonists' time was the printing press, which is an extension of speech. (The telegraph didn't arrive until 1843.) As Benjamin Franklin wrote, "Whoever would overthrow the liberty of a nation must begin by subduing the freeness of speech."[9]

The founding father's wisdom was echoed by Ronald Reagan: "Information is the oxygen of the modern age. It seeps through the walls topped by barbed wire, it wafts across the electrified borders. . . . The Goliath of totalitarianism will be brought down by the David of the microchip."

Reagan was right—initially. American technology combined with a booming economy and savvy economic hardball detonated the collapse of the Soviet Union. The same year the Berlin Wall fell—1989—Tim Berners-Lee invented the World Wide Web, a communication system to share information via the internet. The user-friendly web format reinforced Reagan's vision that digitized information was the ultimate weapon in defense of freedom.

But that was before China and other repressive, authoritarian countries built firewalls to prevent access to banned content and began orchestrating massive technology and data theft across the globe.

Or before Google controlled more than 90 percent of all searches.

Or before Facebook obtained the personal data of 3.6 billion users and then devised information echo chambers to keep users coming back for more tailored, controlled information.

Or before the free content and targeted ads offered by Big Tech helped annihilate competition from newspapers and magazines, effectively shuttering and silencing local journalism.

All these events have been assaults on the marketplace of ideas. Assaults on freedom in order to consolidate a few big companies in the financial marketplace.

These attacks have taken place in broad daylight, often downplayed by our leaders. In 2000, President Bill Clinton joked about China's massive censorship initiative—11 years into the unveiling of the World Wide Web. "Now there's no question China has been trying to crack down on the internet. Good luck!" he said. "That's sort of like trying to nail Jell-O to the wall."[10]

But it's clear now that the joke was on Clinton—and the free world. China practices digital surveillance and digital repression. They are masters of the craft.

And now social media companies like Facebook and Twitter and search companies like Google have been following their lead in a sinister manner. There is a fine line between "content curation" and "censorship." And Big Tech is using it to turn Ronald Reagan's dream of freedom into a repressive nightmare.

TYING IT TOGETHER

Tracing the evolution from the Boston Tea Party to the Declaration of Independence to the Constitution, it becomes clear that the colonists' desire to protect speech and private property rights fueled their protest of monopoly control. As far as our founders were concerned, guaranteeing the right to free speech protected citizens from government interference with an individual's beliefs and ideals.

However, they clearly did not envision a private monopoly with the means to disseminate or quash speech. The printing press facilitated political discourse and debate in the colonies. Local newspapers flourished and accommodated a wide range of political viewpoints. Big Tech has harnessed the World Wide Web, which in many ways is our modern printing press. But, instead of broadening access, Big Tech has done the opposite, silencing certain viewpoints. Imagine the discourse surrounding Thomas Paine's *Common Sense* if it were released today and thought-police sounded an alarm, pronouncing it dangerous: Facebook would fact-check it. Google would de-rank it. And Amazon would de-list it.

The only way Big Tech has managed to act as a silencer is through its monopoly status. It wouldn't be hard to realize the result if the king controlled the printing presses in the colonies. Big Tech is acting as the modern-day king of speech.

Imagine if the East India Company owned a Facebook, YouTube, or Twitter-like platform in the late 1700s. Would it have let John Adams and the Sons of Liberty slam the Tea Act? Would it have distributed videos of the Tea Party or shared links to blog posts of the Declaration of Independence? Or would it have "de-platformed" that content, marking it inflammatory or unsubstantiated? The British Parliament empowered the East India Company; the prime minister was actively working to improve the company balance sheet. It would have been in the company's interest to help the ruling party suppress dissent.

In the era of Big Tech, we don't need to ask "what if" questions. We have seen suppression. As I detailed in the previous chapter, we have seen elected U.S. Senators and Congressmen censored on platforms and suspended from engaging in free speech.

The irony is inescapable: the political leaders who founded America embraced speech to combat repression and economic strangleholds. Now Big Tech is using its platforms to repress speech of political leaders (and others) to maintain its economic and ideological strangleholds. This issue is not one of a few isolated incidents. "Big Tech overwhelmingly censors Republican members of Congress by a rate of 53-to-1 compared to congressional Democrats," found investigative journalist Alec

Schemmel.[11] In addition to well-documented suspensions flagged earlier, Schemmel reports Twitter suspended numerous accounts of conservatives running for office, including the press account of U.S. Senate candidate J. D. Vance. Facebook, meanwhile, bounced an ad by Georgia congressional candidate Dr. Rich McCormick questioning Biden's strategy in Afghanistan.

But Big Tech doesn't have to be overly aggressive. It can simply bury content by adjusting algorithms and changing the relevance assigned to each piece of content. Instead of helping competing concepts surface in the virtual marketplace of ideas, they can effectively silo content they don't like. This "objectionable" content still exists, but you need to know the exact link to find it. Conversely, it can amplify whatever content it likes: a political speech that touts Facebook and Google's civic responsibility? Rank that high! Put it in everyone's feed!

By interfering, ranking, rating, regurgitating, and burying content, Big Tech becomes the all-powerful gatekeeper in the marketplace of ideas.

SAY THAT AGAIN?

Preventing speech because you don't like it is an assault on freedom and the marketplace of ideas. This is why Big Tech's control of vast communications platforms is so ominous. This is also why it is so important to remember what Thomas Jefferson and other founders believed: that monopolies are dangerous and have the power to steamroll over the rights of the many to benefit the few. I laugh when Big Tech apologists suggest that the founders were fine with censorship if it came from monopolistic private actors. Absolutely absurd!

Here's a cautionary tale that doesn't always make it into the history books.

Remember the 1876 presidential election? Me neither. But it turns out this is an extremely relevant election to this discussion because of a behind-the-scenes effort to "fortify" it for one candidate by a two-headed telecom monopoly: Western Union and the Associated Press (AP).

On Election Day 1876, early returns indicated Democrat Samuel Tilden of New York was leading Republican Rutherford B. Hayes of Ohio. Republicans, however, had inside intelligence. Western Union had benefited from Republican-backed westward expansion and laid transmission wires along ever-increasing railroad lines. As such they were in lockstep with the party. They shared Democratic telegrams out of the south with John Reid, the managing editor of the then Republican-leaning *New York Times*. Reid concluded the race was still too close to call. He implored Republican leaders to alert voters that Florida, Louisiana, and South Carolina electoral college votes were up in the air. Alerting operatives was easy—Western Union was backing Hayes all the way and was more than happy to tap out messages to Republican operatives to give no quarter and start lobbying state electoral commissions. Instead of conceding to Tilden, Hayes's handlers claimed victory, which, of course, Tilden disputed.

Access to the telegraph continued to play out during the ensuing stalemate. Western Union had a deal in place to deliver news transmissions from the Associated Press, which controlled the flow of news into and out of newspapers across America. While this relationship raised concerns that the telegraph company was able to influence AP's reporting, where Hayes's election was concerned, the two companies were again very much in sync. In fact, William Henry Smith, the former Republican Ohio secretary of state, and manager of the Associated Press, used his Western Union connections to covertly obtain telegrams between Democrat Tilden supporters, and used his clout in the news organization to distribute legal opinions favoring Hayes while spiking reports on Democratic protests. As Citizen Truth reports, "The collusion between the AP and the Republican party ran so deep that Democrats dubbed the AP the 'Hayesociated Press.'"

In the end, the ensuing stalemate ended with an unofficial deal: House Democrats elected Hayes in return for the removal of northern troops from the South.

This intersection of presidential, electoral, and business history is mind-boggling. It suggests that a tech monopoly allowed the Republican Party to retain power, that the tech company supplied politicians with inside information to influence an election's outcome, that the press abused its own powers of free speech for political purposes, and that Reconstruction ended early because of the resulting electoral impasse and imbroglio, which then facilitated the rise of the Jim Crow power structure that remained in place until the civil rights era.

For decades, Western Union and the AP attracted much hand-wringing and yet avoided regulation. "Western Union's vast system sustained the distribution of uniform AP telegraphic news reports to the press of the entire country," wrote legal historian Menahem Blondheim, "The implications of uniformity and centralized control in this type of information were vastly different from those of standardization in commercial information. Rather than benevolent, these effects could be, and were, recognized as dangerous. Having a single dominant source for political and social information could at best stifle, at worst eliminate, public debate and meaningful public opinion."[12]

In the end, the threat of regulation forced a separation between the two companies. But for Democratic candidate Tilden, it was years too late. Will my colleagues in Congress find this story instructive? Big Tech seems poised to allow history to repeat itself, or worse, outdo itself unless we act.

CHAPTER 3

Repeat Until True

HOW AND WHY BIG TECH FANS THE FLAMES THAT DIVIDE US

Don't you see that the whole aim of Newspeak
is to narrow the range of thought?

GEORGE ORWELL'S *1984*

On October 31, 2016, *Mother Jones* reporter David Corn celebrated Halloween with a chilling article of sensational claims about Russia and Donald Trump. "A Veteran Spy Has Given the FBI Information Alleging a Russian Operation to Cultivate Donald Trump," screamed the headline. Corn did not get to the nitty-gritty right away. Instead, he reported that Democratic Senate minority leader Harry Reid, furious that FBI director James Comey had reopened an investigation into Hillary Clinton's handling of emails, had fired off a letter to the nation's top G-man. "It has become clear that you possess explosive information about close ties and coordination between Donald Trump, his top advisors, and the Russian government," Reid wrote. "The public has a right to know this information."[1]

Using Reid as his cover, Corn reported on the so-called explosive information, relaying that a former Western spy with a background in Russian counterintelligence had provided the FBI with memos asserting Russian sources claimed Putin's "government has for years tried to co-opt and assist Trump." Corn wrote he had read the memos, and he quoted a few snatches: "Russian regime has been cultivating, supporting, and assisting TRUMP for at least 5 years. Aim, endorsed by PUTIN, has been to encourage splits and divisions in western alliance." The memo claimed Trump "and his inner circle have accepted a regular flow of intelligence from the Kremlin, including on his Democratic and other political rivals." But the most disturbing of all the memo assertions was this: Russian intelligence had "compromised" Trump during his visits to Moscow and could "blackmail him."

Corn's Halloween scare story appeared just nine days before the presidential election. Not coincidentally, the *New York Times* issued a story the same day with this headline: "Investigating Donald Trump, F.B.I. Sees No Clear Link to Russia." The accompanying article reported, "Law enforcement officials say that none of the investigations so far have found any conclusive or direct link between Mr. Trump and the Russian government." Instead of exonerating Trump, however, the *Times* piece confirmed Corn's contention that the FBI had received alarming reports, and the *Times'* news of an FBI investigation was equally alarming. Together both articles became source material for more prominent publications and media outlets. With just nine days to go before the presidential election, these ominous allegations began to replicate on Twitter, Facebook, and liberal news sites, like Slate.[2]

Like a toxic, replicating genie unleashed from a bottle, these unsubstantiated rumors about Trump were now out in the open. There was no way to put them back in the bottle and there was no way to stop them from being repeated.

And then, on January 10, 2017, 10 days before Donald Trump took office as the president of the United States, the national media was unofficially suckered into one of the most potent and crippling disinformation operations in history. The shocking stories the spy peddled to Corn and other reporters broke into public view with an article from the mainstream news site Buzzfeed. Here is the first paragraph:

> A dossier making explosive—but unverified—allegations that the Russian government has been "cultivating, supporting and assisting" President-elect Donald Trump for years and gained compromising information about him has been circulating among elected officials, intelligence agents, and journalists for weeks.[3]

The dossier's author was Christopher Steele, a former MI6 British intelligence agent who ran Orbis Business Intelligence, a consulting firm. Steele had done work for the FBI previously. But he assembled his explosive new work at the behest of Glenn Simpson, co-owner of another intelligence research firm, Fusion GPS. Simpson had been hired by a lawyer with Perkins Coie, the law firm representing Hillary Clinton, to conduct opposition research on Trump.

A former *Wall Street Journal* reporter, Simpson previously reported that Trump's campaign chairman Paul Manafort had a history of working with corrupt Russian oligarchs and ignoring laws about filing as a foreign agent. With that information in his back pocket, he knew a narrative, no matter how flimsy, could be concocted about Trump when it came to links with Russia. But he wanted more intel. So, he outsourced the research to Steele.

Steele's first report, memo 080, was dated June 20, 2016. It contained a series of astounding, and unsubstantiated, assertions.

- That Russia had been running a Putin-approved operation "cultivating, supporting and assisting" Trump for "at least 5 years" to divide the West.
- That Russia had provided him with intelligence "on his Democratic and other political rivals."
- That the Russians had offered Trump real estate deals as part of their cultivation campaign.
- And that a "former top Russian intelligence officer" claimed the FSB, Russia's security service, had information to blackmail Trump involving "perverted sexual acts which have been arranged/monitored by the FSB."

Other Steele dossier memos described sinister meetings between Trump team members and Kremlin insiders. One notable tale included how Trump advisor Carter Page met in Moscow with Igor Sechin, head of Russia's state oil firm and a senior

Kremlin Internal Affairs official, and discussed lifting Western sanctions against Russia over Ukraine. Another involved then-Trump lawyer Michael Cohen flying to Prague for a secret meeting with "Kremlin representatives."

In the years since the Steele dossier became public, investigations into the document have been damning. The Department of Justice's 478-page review of the FBI's Crossfire Hurricane investigation into Russiagate interviewed several of Steele's primary sources who disputed Steele's fraudulent work. Sentences like this one are not uncommon where Steele is concerned:

> [T]he Primary Sub-source told the FBI that he/she had not seen Steele's reports until they became public that month, and that he/she made statements indicating that Steele misstated or exaggerated the Primary Sub-source's statements in multiple sections of the reporting.[4]

As far as I can tell, exactly one allegation has been verified: in Steele's memo 086, sources stated that Russia was engaged in state-sponsored cyber operations.

And that's supposed to be news? Pretty much anyone capable of searching the internet would have drawn this conclusion. In 2015, the *Washington Post* ran an article entitled "U.S. Suspects Russia in Hack of Pentagon Computer Network." The only section of Steele's report that was truthful was not ground-shaking stuff. And considering Steele wrote this memo on July 26, four days after Wikileaks had published hacked emails from the Democratic National Committee, you didn't have to be a counterintelligence expert to wonder if Russia might have been involved in that attack.

DISINFORMATION CAMPAIGN

The lies contained in the Steele dossier and the intentional place-ment of this information into the media caused tremendous dam-age to the Trump presidency and our entire federal government. Once the disinformation contained in the dossier was public, it spiraled out of control. The media frenzy was replicated and amplified across social media platforms. Tweets and posts linking to news articles, opinion pieces, TV reports, and hot-take videos exploded. TV talking heads bloviated 24/7. Add it all up, and false suggestions that our president was being influenced and black-mailed by a foreign power became a national crisis.

Hiring campaign chairman Paul Manafort created an open-ing for the false narrative perpetrated by the Clinton campaign because of his connections to Russian oligarchs. Opportunistic political actors also took advantage of certain comments Trump made during his campaign run that provided further ammunition for efforts trying to paint Trump as "pro-Russian."

Those missteps combined with the Steele dossier resulted in a combustible concoction known as misinformation—a very dan-gerous, costly, and modern problem.

Conspiracy theories, whether on the left or right, are hard to debunk or vanquish because it is difficult to disprove some-thing that didn't ever happen in the first place. In other words, if a bank was robbed in broad daylight of a million dollars, we can take witness testimony, use videotape, show transaction records, and prove there was a robbery. But if someone says a foreign gov-ernment has a blackmail tape that nobody has seen of a specific event—how can you prove the event didn't happen or the tape doesn't exist? You can't. The allegation just hangs there. The only "proof" is that the so-called tape hasn't surfaced.

In the digital era, however, no thanks to the business models of Facebook, Twitter, YouTube, and other Big Tech platforms, we now live in echo chambers where information—often misinformation—moves at lightning-fast speed. And the business models of these companies create these silos to increase user activity. Big Tech, then, is instrumental in spreading misinformation.

AMPLIFICATION AND DIVISION

As I noted earlier, Facebook, Twitter, Google, and Apple all use algorithms—a programmed sequence of evaluations—to decide what content appears in each user's feed. Each company wants to optimize the user experience; that is, they want users to interact and like the content the algorithms have chosen. If a user says they like a certain artist's music video, they might show a post related to that same artist. The same goes for political stories. If you read and liked an anti-Trump article about the allegations in the Steele dossier, you would see more of the same. You wouldn't see the stories that showed Steele wasn't a superspy or stories about how other Russia experts, including Steele's own friends, believed he'd been hoodwinked. (Even infamous Trump team critic Fiona Hill, called Steele's dossier a "rabbit hole," saying: "It's very likely that the Russians planted disinformation in and among other information that may have been truthful, because that's exactly, again, the way that they operate.")[5]

The algorithm, then, can shape your worldview, feeding you views or information that may be flawed. The articles or posts might be totally wrong. But social media, in many instances, doesn't care about right or wrong. It cares about clicks, engagement, and obtaining user data that it can sell to advertisers.

Every hot button issue is content fodder for social media. From football players taking a knee to whether masks prevent

COVID to who should win this season of *The Voice*. Express an opinion about any of these issues on social media, or just search on them, and Big Tech will likely serve you up another story related to it. If you click that one, guess what?

You'll get another.

And another.

These content decisions happen very quickly. And posts are monitored to determine which stories are more viral than others. The average tweet on Twitter has a life span of 18 minutes, according to one study. That's because so many people are commenting so often, people with only a few followers rarely have their views amplified by the likes and retweets of other followers. But when the president of the United States is accused of conspiracy or being blackmailed, amplification goes into overdrive; those kinds of controversial tweets replicate with the speed and efficiency of a virile mutant virus.

This system then feeds and broadcasts stories. It funnels information—and *misinformation*—into the marketplace of ideas. This intrusion should be of grave concern to us all especially when the companies are trying to police this information and misinformation and doing a terrible job of discerning the truth. The marketplace of ideas has a real-world impact. Good ideas—the wheel, electricity, the steam engine, the telephone—can have a great impact.

Just look at the creation of the internet. Nobody could have imagined two massive companies, Facebook and Google, would accrue enormous power, and dominate digital communication. The internet and the web were developed to facilitate communication, not allow two corporations to serve as the gatekeepers for the digital transmissions of all written, spoken, and video communication and exploit that communication profit by offering addictive sensational, salacious, and often conspiracy-laden

content and selling ads. Misinformation, on the other hand . . .
Let's look at Russiagate. After all the sensational, alarming allegations, what was achieved? The Mueller Report issued by the Justice Department cost $32 million. No one was indicted. Mueller testified that there was no evidence that the Trump campaign conspired or colluded with Russia during the 2016 campaign. There were countless other investigations and hearings and hand-wringing. It's a safe bet the total bill for Russiagate between the FBI , the Department of Justice, and the special investigations has topped over $100 million.

But the costs have gone deeper. The United States of America became a hugely divided and distrustful place. The FBI was attacked from both sides of the aisle during Russiagate. Ditto the State Department and the Department of Justice. Who wins when our system comes under assault? When real news is buried by the fake news of the day?

While misinformation is obviously a problem, relying on a system where the Government or Big Tech is in charge of determining the truth is not the solution America needs. Free speech must be free. We have to let our citizens absorb news and varying opinions—even if the ideas and "facts" presented are in conflict—and make their own informed decisions. I believe in the American people and their ability to get it right. We are a less free country when social media determines what we should see and share.

Meanwhile, after we've spent millions of dollars and thousands of hours trying to get to the bottom of things, we still have no idea if a Russian disinformation operation actually fed the entire Steele dossier.

But we do know who benefited: Big Tech, which spread the controversy to drive user engagement.

Double Standards

One interesting thing about the Russiagate scandal that is particularly disturbing is that countless inflammatory predictions, accusations, and crackpot theories were posted. Trump would be arrested. Michael Cohen took payoffs in Prague. There really were blackmail videos. Just go look up the Twitter-feed of "citizen-journalist" Louise Mensch (and wife of Metallica's manager) if you want to see paranoid hysteria in action. And yet, through it all, few if any liberal bombthrowers were ever banned. As I document in Chapter 4, Russiagate rumors were breaking before the 2016 election. Things spun totally out of control in January 2017, but there was plenty of sleazy whispering going on from liberal sources.

That said, when stories broke about Hunter Biden's computer in the weeks prior to the 2020 election, major pressure was applied to stop the spread of the *New York Post*'s article about emails that cast Hunter's candidate dad in a negative light. Both Facebook and Twitter actively instituted policies to stop the dissemination of the story!

On February 28, 2019, my colleague Eric Swalwell, a Democrat from California who serves on the House Intelligence Committee, was interviewed by National Public Radio. This interview took place after Trump's former lawyer Michael Cohen, who had pled guilty to campaign finance crimes, testified before the committee, and said he hadn't seen evidence of collusion with Russia. Asked if the president was complicit in Cohen's crimes, Swalwell let fly:

"I'm convinced that there is at least one indictment waiting for President Trump."[6]

In a manner of seconds, Swalwell's quote was everywhere. A leading member of congress, an MSNBC favorite, was predicting a sitting president would be indicted.

It doesn't get more explosive than that.

But evidently, nobody at Facebook and Twitter said, "Hey, wait a minute. Is this really true? Or is this going to hurt Trump?"

There was no special treatment for the conservative in the White House. But the kid gloves came out when it involved the liberal aiming to be the replacement.

I have no problem with Eric Swalwell's right to speak his mind. Let the marketplace of ideas judge the veracity of his statements. Three years later, his comments don't hold water. I point this out as an example of Big Tech's liberal bias and the reason why Big Tech is not suited for its self-anointed role as lord of the truth.

TYING UP "THE INVISIBLE HAND"

In the second half of the eighteenth century, Scottish moral philosopher Adam Smith published two books that contained a three-word phrase: the Invisible Hand. The term appeared in each book only once, but it has evolved into one of the most important, and controversial, concepts in economics. For fans of free market capitalism, it explains untethered, guiding forces of free trade, which Smith believed was governed by the feedback mechanism of the supply-and-demand pricing system. In his view, our market economy is so complex that planning it is impossible, and yet, it serves everyone with mind-boggling efficiency. He cited the example of the manufacture of a laborer's coat, which involved vast segments of society—farms to provide wool, looms and factories, ships—to deliver a single coat. "Without the assistance and co-operation of many thousands," Smith wrote in his 1776 book *An Enquiry into the Nature and Causes of the Wealth of Nations*, "the very meanest person in a civilized country could not be provided, even according to what we very falsely imagine, the easy and simple manner in which he is commonly accommodated."[7]

Smith, in effect, embraces an open marketplace of ideas. The supply and demand of ideas and property, the competition between manufactures and inventors, the buying low and selling high—all of it helps economies and societies grow. He was expressly against any commerce-regulating forces, such as monopolies and collusion.

"People in the same trade seldom meet together even for merriment or diversion, but the conversation ends in a conspiracy against the public and in some contrivance to raise prices," he says.

Economist Friedrich Hayek, whose work *The Road to Serfdom* has influenced generations of conservative and libertarian thinkers, was one of Smith's biggest proponents. "Adam Smith," Hayek wrote in his book *The Fatal Conceit*, "was the first to perceive that we have stumbled upon methods of ordering human economic cooperation that exceed the limits of our knowledge and perception. His 'invisible hand' had perhaps better been described as an invisible or unsurveyable pattern."[8]

Neither of these two influential figures could have foreseen the computer age. And frankly, the engineers who developed the computer couldn't have expected the world Big Tech has carved out. Digital tracking now allows an "invisible or unsurveyable pattern" to be mapped. Privacy may be "masked" by cookies and "virtual private networks," but it can also be unmasked. If you can geo-locate a user to a specific address, how hard is it to mine user data from that area—likes, site visits, videos they watch, the music they listen to—and match it to an actual person?

This is digital surveillance, and it has led to a new paradigm: surveillance capitalism. Harvard Business School professor emerita Shoshana Zuboff, who has written extensively on this subject, defines "surveillance capitalism as the unilateral claiming of private human experience as free raw material for translation into behavioral data. These data are then computed and packaged as prediction products and sold into behavioral futures markets—business customers with a commercial interest in knowing what we will do now, soon, and later."[9]

Zuboff's insights are truly disturbing. If our experiences, thoughts, and actions are being monitored to determine and shape our "behavioral futures," we have a huge problem. Human experience determines the marketplace of ideas, "the invisible hand." If that experience is relentlessly tracked by Big Tech, which it is, our liberty is under assault.

The Harvard professor suggests our human blue-sky nature and quests for profits have led us to this critical point: "We rushed to the internet expecting empowerment, the democratization of knowledge, and help with real problems, but surveillance capitalism really was just too lucrative to resist."

This is a compelling idea. The Big Tech revolution has happened at such high speeds, with so much money invested, America has experienced a sort of digital whiplash. It seems we were sold a future that only the new digital robber barons fully understood. Looking back now, Meta/Facebook CEO Mark Zuckerberg's head of Data Science Team Cameron Marlow bragged to a reporter at the *MIT Technology Review:* "This is the first time the world has seen this scale and quality of data about human communication." He added he was confident his data work would revolutionize the scientific understanding of why people behave as they do and could also help Facebook influence social behavior for its own benefit and that of its advertisers.[10]

When I read this article, my thought was this: Facebook isn't social media, it's *antisocial* media. And anti-American.

Now you can see why I'm concerned about monopolies and free speech.

Big Tech has as much as admitted it wants to control our behavior.

Think Shrink

One of the most chilling effects of Google's information revolution and Facebook's "share what you like even if you don't own it" business model, is that both corporations make billions selling hyper-localized advertising. Thanks to geolocation and data collection, they now boast the ability to display ads based on zip code, income level, previous purchase history, and countless other segments of information about potential customers.

Big Tech was able to do this largely because they got a free pass from Congress. Thanks to heavy lobbying and a new paradigm that cyber-unsavvy members of the House and Senate Capitol didn't fully understand, my colleagues passed Section 230 of the Communications Decency Act. This provision gives internet platforms a free pass to post content they do not own. It expressly states that platforms like Facebook, Twitter, YouTube are not legally liable for users posting trademarked and copyrighted images, music, video, and text on their platforms—even though this illegal content helps build and monetize these companies. At the same time, these Big Tech companies were also consolidating their hold on market share. Google, for example, famously purchased YouTube, the viral video sharing site, giving it the top two sites in terms of internet traffic, and it suppressed the links of rival search engines, such as Foundem.com, a price-comparison engine that vastly outperformed the company's Froogle site.

None of these Big Tech acquisitions and mergers ever faced any oversight or any investigations into the implications of such hyper-personalized, targeted advertising.

Big Tech will swear on a stack of Bibles that targeted advertising is an enhancement for all. That is certainly what Google has claimed. (A Google rep once showed an associate of mine a magazine print ad and dismissed it as "an act of faith.")

But the targeting is enabled by data about users. And that data is amassed via the content users search for, the clicks they make, and the things they post.

If you start to think about Big Tech's "free information" scams too deeply, you might think they have gained control of the marketplace of *bad* ideas. Yes, the digital revolution has achieved many breakthroughs. But Big Tech's power has come with great costs to the average citizen. Forty years ago, most families in America paid monthly bills for a single phone line, cable TV, and maybe a daily newspaper delivery. That was their basic, bottom-line communication expense.

A friend of mine says he now pays more than $300 a month to buy, service, and insure his four-person family's iPhones. By the time he pays off the phones, he fully expects his kids will lobby to upgrade to Apple's newest phone and that the company's inevitable new operating system (or is it a planned obsolescence system?) will likely force him to sign-on for an even heftier new bill. "I'm spending thousands of dollars every year to consume information—at semi-affordable monthly installments so that I don't even notice. Cellphones, cable and internet bills, Netflix, Hulu, Spotify. It's insane. Nobody did this forty years ago. If you needed information, you went to the library. If you wanted music, you went to buy an album, which you could resell. Now we just rent data that we consume on devices destined for landfills."

My friend doesn't expect us to go back in time. But he has a point. The Big Tech monopoly now requires devices and connectivity—things that cost money—to exist and access what now passes for the marketplace of ideas.

And that's a place that should have free admission for all.

CHAPTER 4

A Patented Disgrace

BIG TECH INVENTS THE DESTRUCTION OF INNOVATION

It is better to fail in originality
than to succeed in imitation.

HERMAN MELVILLE

In 2010, the U.S. Patent and Trademark Office (the PTO) granted five-time jump-rope world champion Molly Metz Patent No. 7,789,809.

At the time, Metz thought it was a life-changing event. A car accident had left the Louisville, Colorado, native with critical injuries five years earlier. She worked to regain her fitness, quit her sales job, and set about developing a faster, more precise speed jump rope.

Her favorite device was one of the oldest pieces of exercise equipment in history. The traditional jump rope Metz had grown up with consisted of a handle with a rope inserted into the handle. Together with business partner Paul Borth, she came up with "pivoting-eye-technology"—an enhancement that freed the rope from the handle, eliminating the friction and optimizing the rotation of the rope.

This was a considerable development for several reasons. Metz won her first jump-rope world championship at age 10. Since then, she'd become famous for performing double-unders: the move in which the rope passes beneath the jumper twice during a single bound. On multiple occasions, Metz performed 1,400 unbroken double-unders in 10 minutes. Now, with a more efficiently rotating jump rope, she and the entire jump-rope community would improve their performance.

On a personal level, her Revolution Rope, now the R1 Speed Rope, wasn't just a game-changer; it was a potential life-changer. A new, hi-tech jump rope would be a huge seller in the multi-billion-dollar fitness equipment marketplace.

Two years later, in 2012, Metz and Borth were granted a second patent. Unfortunately, any dreams of financial success and big public stock offerings they might have had quickly evaporated. Instead of jumping for joy, they found themselves jumping through hoops, trying to protect their property.

Big Tech had made that impossible.

PIRACY AND PROFITS

Like thousands of American small manufacturers who have learned the hard way, Metz discovered Amazon was a shoppers' paradise for pirated goods. A search for jump ropes soon yielded results listing multiple knockoffs of the Revolution Rope. Most of these items shipped from China, the piracy capital of the world, and most were priced lower than Metz's product. Adding insult to injury, a decades-old agreement with Universal Postal Union to subsidize fledgling economies reduced the cost of shipping these items from China. The U.S. postal service charged less to ship a product from Beijing to Cleveland than it charged to ship the same item from Denver to Aspen.[1]

Amazon offered zero help. Every supplier of products to Amazon's vast marketplace tells the shopping giant that they have the right to sell their items. That pledge, according to Amazon, absolved the nation's biggest retailer of any responsibility for trafficking pirated goods.

Given that Amazon takes a percentage of every good sold, the company's see-no-evil approach makes perfect sense. The more sales, the more profit. Policing and thereby reducing the number of product offerings is antithetical to its business model.

With pirates infringing on her technology and Amazon happily facilitating sales, Metz contacted Rogue Fitness, an established fitness company, to explore a partnership and showed them her design. According to Metz, Rogue then immediately went rogue, copying the design of the Revolution Rope for its own use.

This outright betrayal was almost the last straw for Metz. She began to withdraw and give up on her dream. But in the end, the jump-rope champion decided to fight back. In 2015, she got a patent lawyer to fight against infringers, and in 2017 she filed suit alleging that since 2011, Rogue Fitness has sold 14 different styles of jump ropes using her patented technology to generate millions of dollars in sales.[2]

A PATENTED DISGRACE

The U.S. Patent and Trademark Office website describes the 2011 America Invents Act (AIA) as a tool "to modernize the U.S. patent system and strengthen America's competitiveness in the global economy."[3]

This would be funny if it wasn't so painfully misguided: the AIA has weakened the patent system and the protection of private property, which is a pillar of the rule of law.

What the America Invents Act invented was a way to allow Big Tech to trample on the ownership rights of patent holders and obtain access to new critical technology while ensuring its own profits and market power.

The so-called beneficial overhaul of patent law didn't just help Big Tech; it hurt our economy, too. The AIA established the Patent Trial and Appeal Board (PTAB), an administrative tribunal with the power to decertify patents. PTAB has been described by a former PTAB judge who summarized the role of the court as a "death squad, kind of killing property rights."[4] There's no "kind of" about it. That's exactly what PTAB has done.

In the first 10 years since the AIA was enacted, nearly 3,000 patents were invalidated—that's an astounding 84 percent of the patents reviewed. When a patent is invalidated, its owner loses control of the "art" that was patented. But the losses don't just end there. If the owner is licensing the patent, the revenues also vanish.

Meanwhile, foreign competitors can now legally poach the patented product or technology and use it for themselves. The foreign pirate gets the privilege of competing against the American inventor who invested time and money in research and design— without spending a cent! The United States now has a law that makes domestic inventors vulnerable not just to foreign competitors but to foreign predators. In June 2020, the U.S. Federal Communications Commission labeled Huawei a national security threat, and since its inception PTAB has considered at least 114 petitions from Huawei seeking to invalidate American inventions.

That's not exactly what I would call "strengthening America's competitiveness in the global economy."

But let's back up. Before we get to the patent changes instigated by the AIA, let's revisit our nation's founding documents.

The U.S. Constitution requires Congress: "To promote the Progress of Science and useful Arts, by securing for limited Times to Authors and Inventors the exclusive Right to their respective Writings and Discoveries."

In his first inaugural address in 1801, Thomas Jefferson described the purpose of government this way: "A wise and frugal government . . . shall restrain men from injuring one another, shall leave them otherwise free to regulate their own pursuits of industry and improvement, and shall not take from the mouth of labor the bread it has earned. This is the sum of good government."

That's pretty clear. Molly Metz, as the Constitution and our founders intended, should be granted the exclusive right to her intellectual property. But the AIA muddies our Constitutional waters.

Just as Section 230 of the Communications Decency Act excuses Big Tech from violating copyright, holding that these corporations are not responsible for illegal content posted on YouTube, Facebook, Twitter, or any other corporate platform, AIA achieved a similar goal. It broadened the rules for denying and striking down patents. Former Alabama senator Jeff Sessions lauded it for allowing "invalid patents that were mistakenly issued by the PTO to be fixed early in their life, before they disrupt an entire industry or result in expensive litigation."[5]

This is a fascinating and disturbing justification. The Constitution doesn't mention anything about industry disruption or expensive litigation.

In fact, for the previous 60 years, Congress barely touched patent law while one new economy-shaking business segment after another exploded. In the sixties, agribusiness enhancements led to amazing efficiencies and increased food production; health-care companies developed a plethora of life-saving medicines and medical devices; energy companies developed horizontal-fracturing technology and energy production increased; and most impressively, the space program paved the way for engineering, aerospace, satellite, and telecommunications revolutions that are still unfolding.

Twenty years before the AIA, our current digital revolution picked up speed and relentless innovations began hitting the marketplace: personal computers, cloud computing, websites, apps, and an endless onslaught of digital devices, from Fitbit to security cameras. But suddenly, with the rise of technology that moves seemingly at the speed of light, with hypercompetitive tech wars between overvalued companies, with China conducting daily raids on American corporations to steal intellectual property, according to PTAB there are too many patents. Really?

What changed about patents? Why the sudden assault to deny them?

One of the arguments for more stringent rules around granting patents is the problem of patent trolls. This, most likely, was what Senator Sessions was alluding to. Patent trolls are companies that don't manufacture anything; they just buy up patents and then launch lawsuits alleging violations. The lawsuits can be seen as a form of racketeering or lawfare in which troll companies weaponize patents, using them to shake down legitimate enterprises, saddling them with uncertainty and crippling legal costs. They are bad guys. And in 2014, the Supreme Court issued a unanimous decision holding that an idea is not patentable—holders must use that idea to make something. This decision was viewed as a huge victory against patent trolling.

But there was another motivating factor to tighten patent protections, one that may seem hard to fathom given the market power and enormous valuations of Big Tech companies. Despite that power and wealth, all the dominant firms are susceptible to competition and even collapse. Tech and intellectual property reporter Michael Shores wrote: "If a group of programmers in their garage could come up with a new set of algorithms that searched more accurately, faster or even in a way that uses less energy, Google could be replaced, or at least have its market dominance threatened. But such a threat only exists if the new market participant is protected by patents. Without patent protection, Google can simply copy the new methods or use its hundreds of billions in offshore cash to buy the new market entrant for less than its full value. Google understood its precarious position as to new and emerging technologies, so it did what any Banana Republic Elite would do—it set out to destroy what it perceived as the real threat: the United States patent system."[6]

DIGITAL POWER AND POLITICAL
PATRONAGE IN ACTION

Money, as a 1980 pop song lyric goes, changes everything. In contemporary politics and business, that goes double.

In 2008, Barack Obama's presidential campaign discovered another powerful force for change: social media. If John F. Kennedy's telegenic charm made him the first TV president, then Obama was the first digital president. His campaign outperformed both Hillary Clinton's machine and John McCain's candidacy in the new media sphere using social media, digital marketing, mobile marketing, and database savvy. Team Obama understood the viral nature of the digital world and used the available tools to articulate his campaign planks, build loyalty, and raise funds.

A lot of funds. And interestingly, a significant amount of those funds came from Silicon Valley. Federal campaign finance reports show that Obama received 12,000 contributions of $200 or more from Silicon Valley to net $9.1 million by May of 2008. In 2012, the size of individual donations climbed even higher as he raised $12.9 million for his campaign and the Democratic Party. According to Politico, 36 e-elites handed over the maximum contribution of $35,800 with seven months to go before the election. The deep-pocket donors included Facebook COO Sheryl Sandberg and Google Chairman Eric Schmidt—"one of four executives from his company maxed out to Obama."[7]

In 2014, all that Google money paid off. Obama nominated one of the company's own—Michelle Lee, Google's former deputy general counsel—as the Patent Office director. At the time of the nomination, Lee was already working for Obama as the Patent Office's deputy director. A woman who had made millions as a Big Tech patent lawyer was now in charge of the world's most potent and vital defender of intellectual property. As influencing operations go, Google and Big Tech scored big.

For small business entrepreneurs, start-up companies, and solitary inventors, the appointment was ominous.

BIG TECH TROLLS AND PATENT TROLLS

There is little argument that patent trolls posed a problem by cynically and unduly disrupting innovation. But what about small business innovators like Molly Metz? She had an idea, and she made something. She wasn't trolling anyone. She was getting trolled.

With the Supreme Court decision and the passage of AIA, the Patent Office now had more ammunition for overturning the patents, including patents that Big Tech and other large businesses found troublesome.

If patent trolling was bad, the AIA had now created an equally damaging and costly new paradigm. Although Lee was nominated as the Patent Office director in 2014, she'd joined the division in 2012, which means she'd been on the inside since the PTAB launched, shaping the world's leading gatekeeper of technology rights. The creation of the PTAB now allows for decertification trolling. A business can file complaint after complaint to overturn a patent. While this is a good weapon against trolls, it is a terrible policy for legitimate inventors.

Just ask Molly Metz. She was successful enforcing her patent rights against several small companies who had been infringing on her patents. The result was a win-win. Metz obtained licensing agreements paying her for her invention and the manufacturer legally produced the best speed jump rope on the market and sold it for a profit. The system worked because Metz's patents were protecting her.

All that changed when Metz filed suit against Rogue Fitness. Rogue responded by petitioning the PTAB to invalidate her patents. If Metz didn't own the patents, they couldn't be found guilty of stealing her registered invention.

For a judicial body that issues life- and world-changing verdicts, often with huge sums of money at stake, the PTAB often seems closer to a kangaroo court than even a small claims court. Unlike a U.S. District Court, PTAB does not allow live testimony or cross-examination. There are no expert witnesses called to testify, and presenting physical evidence is forbidden. Everything is filed on paper. You get the sense that PTAB has never heard the term "due process." Instead, three unaccountable bureaucrats read the petition, study the briefs, and render a decision.

Molly Metz spent six years of her life and tens of thousands of dollars developing and applying for her patents. She had consulted with experts and lawyers, and she even amended her applications to comply with objections. Now, because of the petition to the PTAB, the rights and protections to her inventions that she earned 10 years earlier hung in the balance.

On July 17, 2020, the three-judge PTAB panel composed of patent lawyers with no engineering or mechanical engineering degrees between them, rendered their verdict.

They overturned Molly Metz's two patents.

According to their verdict, the Revolution Rope, which inspired dozens of knockoffs, wasn't revolutionary at all. It was "obvious"—assuming you were familiar with the work of a 1978 German patent application by Gerhard Wolf and a 1979 French patent application by Roger Max Terper.

"A patent claim is unpatentable under 35 U.S.C. §103 if the differences between the claimed subject matter and the prior art are such that the subject matter, as a whole, would have been obvious at the time the invention was made to a person having ordinary skill in the art to which said subject matter pertains," reads part of the 71-page decision, which quotes prior case law: "When a patent claims a structure already known in the prior art that is altered by the mere substitution of one element for another known in the field, the combination must do more than yield a predictable result."[8]

The panel decided that two 40-year-old obscure foreign patents paved the way for the hi-tech jump ropes of Metz and Rogue. Therefore, neither company's products were patent-worthy.

The board also dismissed Metz's assertion that 79 percent of all event winners in national and international jump-rope competitions between 2013 and 2016 used ropes with Metz's enhancement because they saw no proof the winning jump ropes were made by Jump Rope Systems, or a company accused of using the company's patent.

Not long after this decision, I met with Metz. She told me she had lost 90 percent of her income, all of her licensing agreements, and spent what she considered a small fortune trying to protect her innovation and her business. The speed-jumping champion could not jump through all the hoops. "It makes you wonder who the patent system is trying to protect? Inventors or corporations with huge war chests to win every legal battle? It seems rigged against the little guy."

Molly told me she was trying to move on and focus on her fitness center business. Wise move. In 2022, the sages at PTAB denied her appeal.

Obvious Problems

The PTAB's current rules now give enormous leeway for patent removal. The so-called obviousness test can be applied to everything. If examiners find that a "new art" bears a close enough resemblance to an old art, the patent can be revoked because the new art—which nobody might have thought of—is, in retrospect, obvious.

Obviousness is relative. What seems evident to anyone listening to Metz's story is that she showed Rogue her designs, which Rogue then copied and profited from.

Even more problematic—and obvious—is that this precedent is extremely dangerous and anticompetitive in the hands of Big Tech. It is just as dangerous to progress and fairness as patent trolling. Molly Metz's hard work, vision, and innovation were stripped from her by the PTAB in favor of a larger business interest. What is to stop the enormous reserves of Big Tech from challenging and then stealing the innovations of smaller companies? What is to stop the thousands of Chinese pirated products—based on American-made inventions—from flooding Amazon?

The lack of a strong patent system also restricts the marketplace of ideas. This might seem counterintuitive; undoubtedly, some will view the PTAB's relentless overturning of patents under Michelle Lee as liberating innovation. But as Michael Shore notes: "If inventors see the patent system as weak, they will not disclose their inventions, but hide them as trade secrets. This stifles innovation because new inventions that if disclosed could be improved upon are left unavailable."[9]

It's not just the next jump-rope inventor who will be discouraged from innovating. America's economic prosperity and national security is closely linked to home-grown technological advances. Invalidating patents allows our adversaries to copy and produce highly technical innovations that may benefit Big Tech's short-term profits but also empower other countries in the arms race.

Patents have value when they are protected and enforced. Google and its proxy PTO director Lee systematically stripped patent protection. This has devalued patents, which may help Google but hurts the U.S. economy. In 2012, sales of 6,982 patents generated nearly $3 billion in revenue, at an average price of

$422,286. In 2014, the number of patents sold dropped to 2,848, generating $467,731,502 in sales, for an average of $164,232.[10]

Google has now disincentivized invention. The marketplace of ideas is shrinking, not growing. "If an inventor cannot sell his or her invention for a price that supports their time and effort, they will stop inventing," writes Shore. "When they stop inventing, innovation stops."

ELECTION SELECTION

Interestingly, Big Tech likes to change the rules on the fly whenever it suits them. To explain what I'm talking about, let's go back briefly to Obama's tech-savvy 2008 election victory.

One post-election study found that 48 percent of Obama voters received email from his campaign or the Democratic Party. In contrast, only 38 percent of John McCain voters got email from the Republican nominee or his party. Meanwhile, 49 percent of Obama voters shared text messages related to the campaign versus 29 percent of McCain voters.[11] That kind of connectivity and communication clearly paid off in the 2008 election.

After taking it on the chin from Obama twice, the Republican Party improved its digital campaign game in 2016. And nobody was better at messaging than Donald Trump, the first Twitter president. Trump sent nearly 8,000 tweets during his first campaign[12] and over 25,000 messages on the platform during his four years in office.[13]

His running commentary—averaging between 15 and 18 tweets a day—influenced the news cycle on a daily and hourly basis. Regardless of the media's liberal or conservative biases, he controlled the narrative, talking directly to voters in a way that a presidential figure had never done.

Trump used social media in ways that made Big Tech uncomfortable. Liberal Big Tech leaders who donated heavily to Obama and other Democrat leaders did not appreciate Trump's success with the electorate or many of his messages. They didn't like his messages. They developed community standards and then used those standards to silence the most powerful man in America.

So much for free speech.

While Twitter has silenced Trump, the influence of Big Tech on our elections has become shockingly clear. As for all Google's and Facebook's political donations? They are making it back many times over.

Twelve years after Obama rode his digital wave into the Oval Office, the amount of money presidential campaigns spent on digital advertising on just two Big Tech behemoths, Google and Facebook, to reach voters is staggering, with $2.1 billion spent on election ads. Biden for President spent $46,446,358 on Facebook ads from January 1, 2020, to the election, while Trump for President spent $44,135,293. As for Google, Biden's group spent $83,700,800 starting on May 27, 2018, while Trump's team spent $83,428,600.[14]

The campaign expenditures are so close, it almost makes you wonder if Google and Facebook were telling each campaign what the other was spending. But that's just a footnote to the more disturbing story: Big Tech is crafting government policy in its favor, and then profiting off the echo chambers of information it now controls. In other words, candidates can't live without Big Tech.

And if that's the case, perhaps we should be wondering if we can live with them.

LIBERATING INNOVATION

The genie that brought us Big Tech—digital technology—is not going back in the bottle. So it falls to Congress to write the laws to protect our citizens and institutions from monopoly power abuses. I began this chapter by discussing the ways monopoly power destroys innovation. But along the way, I've had to highlight the theft of intellectual property, the gross lack of trademark protection, the devaluation of patents, craven corporate influence peddling, and the intrinsic threat of censorship that arises with information companies becoming too big and powerful.

But since I started with innovation, which drives our economy, our quality of life, our future, let's end by focusing on one way to protect it. We must reform the Patent and Trademark Office and the Patent Trial and Appeal Board (PTAB), so that private property rights can be protected, and innovation can be incentivized.

The Restoring America's Leadership in Innovation Act of 2021 (RALIA) calls for the elimination of the PTAB and rethinking patent protections. Other proposed legislation reforms PTAB by inserting limited due process protections in the adjudicative process. Indeed, why award a patent and then allow eternal challenges to it? Instead of rewarding an inventor with ownership, it rewards the richest and most litigious (and their lawyers) who can afford to assault and destabilize a patent. If necessity is the mother of invention, then the Patent Office's PTAB is the undertaker. Let's bury it or at least put it on life support—and help inventors protect themselves from Big Tech lawfare assaults and feed the marketplace with new ideas.

CHAPTER 5

The Puppet Show

HOW BIG TECH CENSORS AND DEVALUES THE PRESS

.

Give all the power to the many, they will oppress the few.
Give all the power to the few, they will oppress the many.

ALEXANDER HAMILTON

When Weifeng Zhong left mainland China to attend graduate school at the University of Hong Kong in 2006, he spotted a startling sight on his first visit to the campus: an eight-meter-high bloodred sculpture. The work, entitled *Pillar of Shame*, was a depiction of 50 dead and mangled bodies, piled on one another, with agony etched on their faces. At first glimpse, Weifeng thought the crimson tower was an eyesore, but one that ignited his curiosity. He moved in for a closer look.

At the base of the sculpture, Danish artist Jens Galschiøt had inscribed three lines of text in both Chinese and English: "The Tiananmen Massacre," "June 4th 1989," and "The old cannot kill the young forever."

Weifeng was mystified. *The Tiananmen Massacre?* he wondered. *What massacre is it talking about? Nobody died in the Tiananmen Square protests.*

The Tiananmen Square rebellion—and subsequent massacre—transfixed the world in the spring of 1989. In the wake of the April death of politician Hu Yaobang, students and other activists began to gather at the square—a popular Beijing destination—and peacefully protest for greater freedom of speech, democratic reform, and the resignation of government leaders. On June 4, after weeks of protests, troops stormed the square, firing at protesters and conducting mass arrests. The "official" government body count was 200 dead; other sources have placed the death toll as high as 10,000.

More than three decades later, the chilling standoff of a single, unknown Chinese man, a shopping bag in each hand, standing in front of a phalanx of tanks at the square, ranks among the most iconic images of the twentieth century.

Weifeng, who was six years old at the time of the tragic uprising, had never seen that image.

He had no idea that the Chinese Communist Party (CCP) had ordered violent action suppression. Or that as many as 10,000 peaceful Chinese citizens had been murdered by their own troops.

Weifeng, who now lives in the United States, was a guest on my *Shooting Straight* podcast. More than 15 years after encountering the *Pillar of Shame*, he told me the event still leaves him stunned. "I went to the library and walked past a whole shelf of books about one subject only, the Tiananmen Square Massacre. I remember staring at those books, thinking, 'Wait a second! Who died?' Nobody died, according to what I was told. So, I was in shock. That's the shock of my life."

Weifeng began reading the books. He watched documentaries. It was difficult to process.

"I watched with excruciating pain, not just because it was a tragedy, but also because I was shocked by how effective propaganda was when I was growing up."

That pain inspired Weifeng to forge a new path. He wanted to make a career analyzing the propaganda orchestrated by the CCP.

"In the past, when I was in mainland China, I was a puppet in a puppet show. Now I'm in the audience. I'm watching the puppet show. Now I'm able to see what the puppeteer is doing to the puppet."

While I found part of Weifeng's journey inspirational—another transformational American immigrant story—I also found it terrifying and instructive. His ignorance of Tiananmen Square demonstrated the frightening power of censorship of the press. There is no bill of rights in China; the interests of the state—of the CCP—are paramount. A deadly, cruel, event orchestrated by the Chinese government had been airbrushed out of history by the very party that was responsible for it.

Then the story of Tiananmen Square's squelching is an instructive story about big government, freedom of the press, and the censorship of and influence on that press. To understand how these issues intersect with Big Tech, I want to return to a central idea of this book: that free speech depends on free markets and free markets depend on free speech.[1] What's the connection? Well, the short version is that Big Tech's monopoly power threatens these two pillars of America.

DEPRESSING THE POWER OF THE PRESS

In the opening chapter of this book, I documented how Facebook and Twitter actively engaged in suppressing a potentially damaging story in the *New York Post* about Hunter Biden prior to the 2020 presidential election. Although these social media platforms obviously don't have the authoritarian power of the CCP (yet, anyway), there is an obvious parallel: both were engaged in censorship and gatekeeping in the marketplace of news.

It is fitting, then, to delve into the history of the *New York Post*, which has contributed to that marketplace for 220 years. The oldest continuing running daily paper in America, it launched in 1801 as the *New-York Evening Post*, the brainchild of Alexander Hamilton, who wrangled investments with the backing of fellow members of the Federalist Party. Hamilton wanted a paper to cover and counter newly elected president Thomas Jefferson, who didn't share the Federalists' beliefs in a strong central government. Hamilton's paper was heavy on business and shipping news and occasional crusades.

Over the decades, the paper has evolved with the times and provided some of the most celebrated journalists in the country.

Editor William Cullen Bryant, a poet and abolitionist, turned the paper into a profitable powerhouse that was noted for its liberal politics. In fact, English philosopher John Stuart Mill, a close colleague of "invisible hand" theorist Adam Smith, singled out the paper for praise in 1864.

After Bryant's reign, the paper ran through a slew of owners on both sides of the political spectrum and often struggled to make a profit. But in 1934, millionaire Dorothy Schiff took over the paper and ran it for 42 years. During that time, it became a hot bed of liberal Democrat voices like Eleanor Roosevelt, Max Lerner, Murray Kempton. It also launched the careers of beloved writers like Pete Hamill and Nora Ephron, the future director of *When Harry Met Sally* and *Heartburn* (and ex-wife of Watergate reporter Carl Bernstein).

In 1976, Rupert Murdoch, then the king of British tabloids, bought the *Post* from Dorothy Schiff. Under new ownership, the paper grew famous for loud, sensationalist celebrity coverage, but it also became a must-read for conservatives, with columnists like Thomas Sowell, Evans & Novack, and Michael Goodwin, and reporters who are unafraid of targeting the swamp-like doings of Bill and Hillary Clinton, the Obama administration, and most recently the Bidens.

Murdoch was forced to sell the *Post* when he purchased the Fox network, but when the paper faced bankruptcy under its new owner, Peter Kalikow, the Federal Trade Commission (FTC) allowed him to buy the paper back.

The *Post*'s history is fascinating to me because, even though it has had many owners with various political perspectives, and even though it was often in precarious financial straits, it has endured and has continually fueled the marketplace of ideas with compelling columnists and reporters who aren't afraid to blow whistles.

With the arrival of Big Tech, however, whistleblowing, as we saw with the *Post*'s Hunter Biden story, can be muted.

And silenced.

A CREATION AND ELIMINATION STORY

The first brilliant idea Google founders Sergey Brin and Larry Page had was to "crawl"—and trawl—the World Wide Web. Using spiders, it would copy what it found, create vast databases of web content, and then "index" that content, evaluating it for relevancy against search queries. The logic rules driving those relevance evaluations are known as the Google algorithm.

That algorithm then is used to return search results. As people in the web business say, Google is vital when it comes to "surfacing content." In other words, if a reporter writes a story about, say Popsicles, he or she will hope Google indexes the story and that it scores high in the algorithm so that when anyone searches "Popsicles," the story will appear on the search results page.

Google then launched its business using other sites' content. And one of the primary reliable content creators is newspapers. I'm tempted to say the amount of the free content newspapers have provided over the years for Google is incalculable. But on second thought, since Google indexes and tracks everything, it probably knows.

In 2018, Google earned an estimated $4.7 billion by including links to newspaper articles in its search results, according to a study by the News Media Alliance. Google doesn't produce any original news content; earnings are primarily generated by selling targeted ads tied to its search results. This means the content of news articles—searchable words and images—is directly connected to driving ad sales.[2]

According to the study, news articles represented 40 percent of the links on search results. Despite that, Google did not pay a single cent to newspaper publishers for displaying their content and providing links to their stories. Does this model of showing someone else's original content sound familiar? It is at the heart and soul of Big Tech, which wants to make everyone's content free while it harvests money from other related channels. In the case of search results, Google doesn't have to rely on Section 230 of the Communications Decency Act—the law that both YouTube and Facebook were built on, which says sites aren't responsible for the presence of content that violates trademark or copyright regulations. Instead, displaying an article's headline and initial content is allowable under fair-use laws.

Since 2008, when Google executive Marissa Mayer estimated news was worth $100 million to the company, Google profits have skyrocketed. Alphabet, the company that now owns Google, reported revenues of $75 billion in Q4 of 2021, of which $62 billion was generated by Google advertising.[3]

You don't have to be a Mensa member or Big Tech data scientist to realize that Google has increased its value by crawling the web, indexing news content into relevant buckets, giving those results to users and then tracking what the user clicked on. Or that it has simultaneously devalued the work of the newspapers it linked to because although Google didn't research or write the article, it "found" it for the consumer free of charge.

Google, Facebook, and Twitter maintain that sharing links to news articles helps newspapers. The evidence to support that claim is slim to none. Newspaper ad revenue shriveled from $45 billion in 2007 to $16 billion in 2017, while Google and Facebook valuations climbed. During that same period, the number of newspaper employees shrank by 26 percent.[4]

Too Little Too Late

At the end of 2021 in Australia, Google had an estimated 95 percent market share in search and a 96 percent market share in search advertising. Around the same time and perhaps because of Google's dominance, legislation surfaced in Parliament forcing Big Tech to pay for the news posted in search results or site feeds. Google quickly cut a deal to pay newspapers and, reportedly, share some data. Reuters, one of the leading global news services, also signed an agreement with the search behemoth.

Sharing a slice of Big Tech's vast wealth will not undo Big Tech's damage to the newspaper industry and civil, informed society.

By optimizing and selling its ad vision, Big Tech has stripped local newspapers and magazines of the advertising dollars these community institutions rely on to survive. The drop in ad revenue has proved lethal. More than 60 American dailies and 1,700 weeklies have closed since 2004.[5]

The papers that have survived have seen their news-gathering budgets slashed. This effectively silences speech within communities and shrinks the marketplace of ideas at a grassroots level.

"When you lose a small daily or a weekly, you lose the journalist who was gonna show up at your school board meeting, your planning board meeting, your county commissioner meeting," says Penny Abernathy, a professor at University of North Carolina who tracks paper closures and worries that the shutdowns are creating "news deserts."[6]

Big Tech believers will undoubtedly dismiss this point with claims that they have done the opposite. The marketplace of ideas has expanded, goes the argument, because everyone has a voice on social media. This assertion, too, is disingenuous. We know the average social media post effectively evaporates after 18 minutes or less. A letter to the editor of the town weekly lasted a minimum of seven days. You didn't need a device that cost hundreds of dollars to read that letter—or pay an internet service provider hundreds more for access to the web, either—because many local papers were free.

While most social media posts vanish quickly, there is also an important and vast difference between the average social media post and a newspaper article. News stories are generally written by journalists who conduct research and, ideally, verify facts. Usually, there is an editor or two who also vet the article. That isn't to say all news articles are 100 percent true—sometimes they are indeed slanted—but they are mostly based on fact. On social media, a vast proportion of posts consists of opinion, conjecture, and speculation. Sometimes these posts pose as fact, and just as dangerously, they are read as fact.

AD MONOPOLY

Big Tech platforms have grown with the rise of cellphones and other digital devices. There is no question these increases have broadened the reach of free, web-based news content. "We have an exponentially larger audience than we ever had," Danielle Coffey of the News Media Alliance told CBS News. "Consumers come to Google, they go to Facebook, and they get our news content. And when that happens, we're stripped of a large portion of advertising revenue."[7]

Speaking of digital ad revenue, depending on your point of view, Google is either brilliantly and indestructibly vertically integrated, or it has a monopoly in that space. In 2007, it purchased DoubleClick, a market-leading digital advertising firm that owned a platform allowing web publishers to sell their ad space and a platform that managed advertising bids. In a single deal, the search engine with a dominant market share now owned the dominant ad-serving and ad-buying platforms. As I will explore in more detail in a later chapter, Google not only charges transaction fees to both ad buyers and sellers, but it also has access to the data surrounding every click in its enormous ad universe.[8]

It was a brilliant business move. One that, in retrospect, should have been challenged at the time of the acquisition. The FTC, however, gave the deal a pass. It approved the DoubleClick deal in a 4-to-1 vote, and at least one of the members who supported the merger now wants his vote back. "If I knew in 2007 what I know now," former Federal Communications Commission (FCC) member William Kovacic told the *New York Times*, "I would have voted to challenge the DoubleClick acquisition."[9]

A second former FCC commissioner, speaking anonymously, admitted to the paper that no one imagined the power that Big Tech players like Google, Facebook, and Amazon would amass.

No one, that is, except the strategists at those companies.

DO NOT PASS GO

The destruction of the newspaper industry is tied to the ubiquity of Google search and the power of its ad platform. If Google can monopolize searches—on the web, from within apps, on browsers such as Chrome and Safari, and from digital devices—then it not only dominates advertising on the web, it also controls the flow of news information in the digital world.

In 2020, the U.S. Department of Justice (DOJ) finally decided enough was enough. Joined by 11 state attorneys general, the DOJ filed a civil antitrust lawsuit "to stop Google from unlawfully maintaining monopolies through anticompetitive and exclusionary practices in the search and search advertising markets and to remedy the competitive harms."

The 64-page complaint chronicles a litany of anticompetitive abuses by Google. The DOJ charges Google uses "monopoly profits to buy preferential treatment for its search engine on devices, web browsers, and other search access points, creating a continuous and self-reinforcing cycle of monopolization."[10]

Among the most damning charges are that Google seeks prime real estate on devices, apps, and websites to corner the search engine market. "Google pays billions of dollars each year to distributors—including popular-device manufacturers such as Apple, LG, Motorola, and Samsung; major U.S. wireless carriers such as AT&T, T-Mobile, and Verizon; and browser developers such as Mozilla, Opera, and UCWeb—to secure default status for its general search engine and, in many cases, to specifically prohibit Google's counterparties from dealing with Google's competitors. Some of these agreements also require distributors to take a bundle of Google apps, including its search apps, and feature them on devices in prime positions where consumers are most likely to start their internet searches."[11]

One relationship in particular seems to have infuriated the DOJ, and it's easy to see why. The search giant struck "long-term agreements with Apple that require Google to be the default—and de facto exclusive—general search engine on Apple's popular Safari browser and other Apple search tools." In other words, Google, which owns the most popular mobile operating system, did a deal with the owner of the second most popular mobile operating system to become the default search engine on Apple's phones.

WHY THIS MATTERS

The ideals that propelled the founding of America are rooted in preventing absolute power—in government or in business. Federalism, separation of powers, and many provisions of the Bill of Rights help prevent the government from getting too big and infringing on our freedom of speech. How do we prevent Big Tech from getting too big and infringing on our freedom of speech or subtly censoring or filtering it?

It can't be overstressed: free speech is vitally important because it spurs the marketplace of ideas, which spurs the economic marketplace. The two marketplaces are integrally linked. But if Google and Apple hate every patent they don't own or candidates whose ideas don't jibe with policies that help consolidate Big Tech's market power, then it can silence or suppress speech simply by "adjusting" the relevance results returned by that "default" search engine.

The collapse of local newspapers doesn't end with the loss of jobs and the elimination of important big news stories. The small stories that unite communities disappear, too. The coverage of church events, town hall meetings, local fairs, and local team sports evaporates. These types of stories help communities define and cement themselves. They are a bonding agent. And when you see a newspaper—an actual, physical, collated group of pages, as opposed to website—it is a visual confirmation that a defined area—a town—exists, not just as a dot on a map, but as group of people living, playing, sharing, helping, and governing.

Elaine Godfrey, an *Atlantic* reporter, tracked the demise of *The Hawk Eye* in Burlington, Iowa, the oldest paper in the state. She also tracked the local groups on Facebook, one with 16,000 members.

"It's ostensibly a site for sharing news about the community, but the page is chaotic. Scroll down and you'll find mug shots of tattooed men alongside pictures of missing dogs, ads for rib eyes at Fareway, and comments accidentally made in all caps," Godfrey reports, adding: "The pages can be a useful resource, and a good source of community jokes and gossip. But speculation and rumor run rampant. A member might ask about a new building going up in town, and someone will guess that it will be an Olive Garden. It never is."[12]

What fills the vacuum of a local newspaper, then? Some accurate information, sure. But plenty of rumors, lies, and misinformation. Instead of nuanced local reporting, America's smaller towns become fly-over localities. The only news is from somewhere else. In these instances, social media isn't liberating—it's disempowering. The press is supposed to be empowering.

For more than 200 years, newspapers have been a beacon of free speech in America. There have been all kinds of publications through the generations, many spewing ideas that we now consider vile or ignorant. The marketplace has consigned them to the dustbin of history. But just as John Dickinson demonstrated by printing a Philadelphia broadsheet ripping into the Tea Act and the monopoly of the East India Company, the free press has blown the whistle on wrongdoing. It has helped shape opinions, disseminate ideas from the marketplace, and contribute to our ideas of liberty.

When you look at the antitrust charges against Google, the implications are frightening. By targeting search and owning the market, its influence is chilling. It is precisely the kind of power our founding fathers were against.

That market power is also out-and-out unfair, according to the European Union. In 2017, regulators at the European Commission in Brussels hit Google with a $2.7 billion fine for giving its shopping comparison engine Froogle special preference over other sites in its search results.

"Google's strategy for its comparison-shopping service wasn't just about attracting customers," said Commissioner Margrethe Vestager. "It wasn't just about making its product better than its rivals. Google has abused its market dominance in its search engine by promoting its own shopping comparison site in its search results and demoting its competitors."[13]

The search engine was acting as a gatekeeper of the marketplace.

In 2022, the connections between this corporate entity and our entire society are so entrenched that they threaten our democracy. Its dominance is now anticompetitive.

As we saw in the last chapter, Google and Facebook rake in hundreds of millions of dollars in political ads placed by candidates and campaigns. They control the dissemination of articles about those candidates and campaigns, Google determines which will be seen first and Facebook which will be shared at all. Meanwhile, on YouTube search results determine what clips will be offered to searchers first. It also determines what paid-for speech is heard first—as its most prominent links are sold to bidders.

We can't dump our web browsers or cellphones into the Boston Bay—we now use them to access the rest of our lives, our bank accounts, our address books, our diaries, and all our entertainment. Google's concentrated power didn't just "happen." The company business model is based on a slyly deceptive exchange: it offers a public service—listing relevant information that exists on the web—in exchange for showing users targeted ads while also tracking their searches and clicks, essentially gathering data that will yield more clicks and more ad revenue, which is generated by optimizing that data.

Big Tech's collusion also helps consolidate power. The DOJ's antitrust charges are alarming, particularly its assertion that Apple agreed to make Google the default search engine on the popular Safari browser that is embedded on all Apple devices.

Both companies insist this was business as usual. But then that's the point: America has lost the plot when unfair becomes usual.

While the DOJ is pursuing antitrust charges against Big Tech, other antitrust laws have kept America's newspapers from defending themselves. As crazy as it sounds, newspapers and other news providers are prohibited by the antitrust laws from banning together and negotiating a deal with the monopoly tech platforms. That, apparently, would be unfair—to the monopolies!

You can't make this stuff up.

ECONOMIC POWER AND TRUTH

Weifeng Zhong now spends his days monitoring the vast amounts of propaganda spewed forth from the Chinese Communist Party. His goal is to analyze public statements to detect policy shifts from Beijing before they actually happen. Talking to him on my podcast he touched on what he regards as the key to China's leverage in the global marketplace: economic power.

"China has a lot of economic leverage, that could make some values that we highly treasure tradable," he said, noting that even for America's allies "everything is tradable. Everything is sellable, so to speak, right? Think of free media in some of those developing but democratic countries. 'Do you want to offer up a space to run some rosy Chinese stories?' Some will say yes, because they are resource poor."

It is a sobering thought to compare Big Tech monopolies, which monitor our digital world and filter news, to China, a totalitarian state with no regard for basic human rights. But the similarities are there. With Big Tech monopolies controlling what we see and say, consumers get neither the ability to express themselves freely nor the power to choose different ideas in the market.

Ensuring our free press has the financial strength to operate and fulfill its role as an unimpeded publisher of information is critical to keeping the marketplace of ideas open. To counteract the erosion of the press's financial stability, initiated in part by Google and Facebook, we need to allow the free press to defend itself. One legislative idea should provide some relief by lifting antitrust restrictions against small media companies banding together to negotiate. This will let publishers negotiate and collect from the digital platforms that have greatly profited from news organizations' work. The proposal provides a four-year safe harbor window to hammer out terms to recoup advertising and subscriptions from digital giants who used news articles to gather data.

Google and Facebook don't have any incentive to negotiate with deserving members of the media because as it currently stands, the monopolies get their content for free and receive all the revenue. The only leverage that small newspapers can have is to band together and deny Google and Facebook a large quantity of original content. But I hope a legislative solution can be crafted that allows a more even negotiating position. As Weifeng Zhong's story suggests, the future of American liberty depends on it.

CHAPTER 6

A Poisoned Apple

SELF-PREFERENCING AND THE APP STORE'S RESTRICTIVE POWER

_____

There is no friendship in trade.

CORNELIUS VANDERBILT

It was one of those icy 19°F January mornings in 2020. I was driving from my home to Boulder, Colorado, and I was not happy. The Antitrust Subcommittee had scheduled our fifth hearing at the University of Colorado Law School. I love every corner of my home state, but we refer to the People's Republic of Boulder as 28 square miles surrounded by reality. It goes without saying, I don't really fit in.

And as the only Republican member of the committee attending this hearing, I expected to feel even more like the odd man out. So, I decided to spend as little time at the hearing as possible.

But the primary reason for my grumpiness was that I thought the hearing was a waste of time. I was expected to attend this subcommittee hearing—"Competitors in the Digital Economy" was the theme of the day—because it was taking place in my backyard. But I didn't expect to learn anything. Heck, I already knew what needed to happen. I was a fiscally responsible, free-market-loving conservative. *The best way to deal with problems in the marketplace is to let the market address the problems.* That was—and is—my economic mantra, and many of my Republican colleagues swore by it as well.

During my drive, I outlined in my head a few points I wanted to make in my opening remarks. Before I heard a word from the witnesses, I issued a preemptive warning of sorts, laying out the three principles that I wanted my colleagues to bear in mind. "First, innovation and competition in the tech center have produced enormous value for consumers," I said. "We should not forget about those benefits as we consider the current state of competition in the tech center."

"Second, any legislative proposals that emerge from our inquiry should be consistent with maintaining a free and competitive marketplace. Proposals to construct broad new regulatory regimes must be viewed with caution. Experience has shown that burdensome regulations often miss the mark."

"Third, big is not necessarily bad. Antitrust laws do not exist to punish success but to promote competition."[1]

That was my mind-set as the hearing kicked off.

A panel of four speakers presented brief statements, each less than five minutes long. The stories they told were about their companies—successful, market-leading start-ups, including Sonos, the publicly traded leading manufacturer of smart music systems; Popsockets, the innovative phone accessories leader; Basecamp, a software service platform like Slack; and Tile, a pioneer in location devices and software. Each of their tales shared a common theme. They began as small, dynamic companies. They burst on the marketplace and showed explosive growth, often with the help of Big Tech. And then, suddenly, they hit roadblocks—anticompetitive tactics casually, cruelly, and calculatedly employed by those same Big Tech companies. Why? Because the monopoly executives had seen the innovative work and the market that these start-up companies had created, and they now wanted to own that market for themselves. In short, the witnesses described a transformed marketplace that mutated from open and creative to hostile and domineering.

The most moving aspect of their presentations was something I never expected from an unrelated group of incredibly dynamic, innovative, independent, and successful Americans. They shared a send of powerlessness. Their testimony was a plea for help.

While all the participants were compelling, Tile executive Kristen Daru's testimony truly resonated. Her company is a market leader in a useful product—tracking devices. Users attached a "Tile" to their belongings and when those items—backpacks, watches, computers—are misplaced or lost, the Tile app installed on a phone helps locate them. The company grew quickly, with many users buying Tile tracking devices at Apple stores and using the app on their iPhones.

Tile's world began to change in April 2019. Apple developed a competing, "knock-off" device and soon thereafter informed the company that they would no longer carry the Tile device in Apple retail stores. Evidently, thanks to Tile, Apple discovered the potential of the personal item tracking device, copied the concept, and used its market power to crush its former business partner.

Daru explained, "Apple's FindMy app competes with Tile. More importantly, FindMy is installed natively on Apple hardware to the exclusion of all competing apps, and it cannot be deleted"—unlike Tile.

She went on: "Also with iOS 13, which released last September, Apple made it more difficult for our customers to enable their Tile devices by burying the required permissions deep within the consumer setting"—unlike FindMy location settings, which "are activated by a single selection during operating system installation."

"Once our customers figure out how to enable our devices also with iOS 13 Apple started surfacing these reminders and prompts to our customers to encourage them to turn it off, but Apple does not serve any reminders for its customers to turn off FindMy's location permissions."

Her testimony detailed aggressive bait-and-switch online ads that looked like Tile promotions but took them to FindMy in the App Store. She also noted that because Tile had been sold in both

Apple retail stores and the App Store, Apple had access to "competitively sensitive information including who our customers are, our retail margins our subscription rates."

Listening to these details made under oath from a digital veteran—Daru spent 10 years directing digital policy at the video games giant Electronic Arts—was riveting. And it led to one disturbing conclusion: Apple had launched a highly orchestrated anticompetitive bombardment against Tile.

Just as arresting, I heard Daru echoing my own thoughts about open marketplaces, accusing Apple of "acting as a gatekeeper of third-party access to permissions and technology in ways that favor its own interests."

In addition to her marketplace concerns, she also talked about my other favorite metaphor for open, flourishing economies and democratic societies: the playing field. I knew Apple owned the entire commercial iOS ecosystem—the hardware, the software, and the App Store. But Daru exposed how Apple abused that ownership by changing the operating system without notice so that competitors' products wouldn't sync with the new Apple software. Tile was competing on a playing field while Apple was actively changing the landscape.

Boom! Daru's testimony challenged my beliefs about economic fairness and my cynicism about burdensome regulation. Her testimony, along with others before, made it crystal clear that Big Tech was using its vast market power as a weapon. But I wondered how to deal with this immense inequity.

"What is the answer . . . should Apple not be allowed to create competitive products?" I asked Daru during my questioning time. "Should Apple be restricted in its pricing or the way it regulates products in the App Store? What is it that you would like to see?"

Her answer was equal parts transparency, visibility, and access to permissions and critical technology. "We need advance notice of changes to the iOS that have a meaningful impact on our business, and we need rules that are applied consistently across the board to everyone in the ecosystem, including Apple," she said, adding, "We have to watch how we can address these inequities because ultimately the future of competition in the United States depends on it."

"So, you're all right with Apple creating a competitive product?" I asked. "It's the way that they distinguish or discriminate between the two products that is the problem."

"Absolutely. We welcome fair competition, but it has to be *fair*. We're seeing time and again, Apple using its dominant market power and engaging in practices that put us at a competitive disadvantage . . . and it's those types of unfair practices that we need to curb."

Competition and unfairness were themes that had come up repeatedly with the other testimony. And I realized that this was the first time I was hearing the other side of the story. Big Tech spends tens of millions of dollars whispering in the ears of Capitol Hill policy makers.

My last question of the hearing was directed to the former philosophy professor and inventor of Popsockets, David Barnett, and foretold my changing attitudes about the Big Tech monopolies: "You can't compete unless you're making a profit. If there is a company or a country that stifles your ability to make a profit we will see less and less innovation in this country and . . . as we see less and less innovation in this country, we will not be a dominant player in the world because we don't manufacture at a lower cost, we innovate at a higher rate. Agree?"

"I fully agree in each instance," Barnett said. "If government can take action that will help protect innovation without doing a greater harm, government should do that."

The hearing ended and I joined our witnesses over lunch. I asked each of them privately if they felt their stories were unique.

The answer was a resounding no. They recounted one story after another about inventors and entrepreneurs—imaginative, ingenious Americans who rolled up their sleeves and got things done—who grew terrific businesses only to run into brick walls erected by Big Tech.

As I noted in the first chapter of this book, this was a watershed moment for me, an event that left me thinking, *Holy cow, we've got a major crisis unfolding and the rest of the country thinks everything is A-Okay.*

I drove back home feeling like I'd taken a complete 180-degree turn from my grumpy morning mood. Now I was outraged and, to be honest, frightened. The examples of monopoly power and abuse I'd heard were extremely disturbing, but there were two other things that I found chilling. First, the current state of play was challenging my hardwired assumption about free and open markets intrinsically fixing themselves. I still believed that. What I no longer believed was that the market was free and open where Big Tech was concerned. And that led to my second concern: Where monopolies tread, censorship and the threat to liberty is sure to follow. And this was doubly worrying because Big Tech is laser-focused on the software and devices that allow us to communicate and share bold thinking in the marketplace of ideas.

And that makes Big Tech's unbridled power a threat to democracy—they must not have the power to act as gatekeepers of the idea marketplace.

I am a hard-core conservative. I believe in government staying out of the way whenever possible. But what I'd heard in Boulder could not stay in Boulder.

Since I took office, I have sometimes wondered what I am doing in Congress. I don't mean that I was heading to Florida, took a wrong turn and found myself in Washington, D.C. No, I worked hard to win elections, solve problems for Americans, and represent my constituents' concerns. But I came to D.C. to reduce federal spending, promote local solutions, and cut the tax and regulatory burden on small businesses. With a few exceptions, Congress has done the opposite. I realized after the Boulder hearing that leveling the playing field for innovators in their battle against monopolies was a righteous calling and energized my purpose for being in Congress.

I also saw that the Big Tech fight was vital to protecting our nation. I knew I had to act and alert my colleagues about the insidious enemy—one made up of companies the average American might love and admire—that now touched every facet of society and was in a position to not just broadcast ideas but silence the ones they didn't like.

The outrage and trepidation I felt were replaced by better energy—I was driven and inspired. I knew what I had to do. As I motored home, something strange happened. I was excited to get back to Washington. To sound the alarm.

WARPING THE PLAYING FIELD

Apple's tactic of claiming it is protecting user security and then using that policy to hurt the products of rival apps is effective. As the owner of iOS, it makes the rules. As the second biggest app store in the world (Google Play has more apps and more downloads), app makers need to play by its rules. But Apple and Google shouldn't be able to promote and feature their own products over the competing products made by rival businesses. This self-preferencing discriminates and effectively closes the market.

Apple may hide behind claims of protecting customer privacy. But the sequence of events leading up to the rollout of AirTags, the location device that competes against Tile, makes the company's behavior petty and cut-and-dried. Apple clearly self-preferences its own apps and services to artificially restrain competition from Tile.

Apple, of course, has admitted no wrongdoing. In a statement by Kyle Andeer, Apple's VP of Corporate Law and chief compliance officer, he didn't breathe a word about self-preferencing. "I think the App Store's real success is the opportunity it has created for software developers to build and distribute all of those apps in the first place," he said, adding, "It truly revolutionized software distribution." Then he shifted gears, launching into Apple's security mantras, "We review every app in the App Store to make sure it meets standards for privacy, safety, security, and performance."

Surcharge or Unfair Tax?

Apple does something else when it reviews products for App Store placement. It decides whether to slap a 30 percent surcharge on any content or service app that offers paid subscriptions or in-app purchases—like Spotify's music streaming options or Match's dating app, or the hugely popular *Fortnite* video game app.

The price for selling an app in Apple's closed iOS ecosystem is steep, unrelenting, and often self-preferencing and anticompetitive. In February 2011, Apple eliminated payment choices, required all apps to use its billing system, and prohibited them from offering links to external ways to pay.[2] According to Spotify, it was "the first of many moves from Apple that would make it harder and harder for our fans to upgrade to Premium."[3]

As I reported earlier, Apple acquired Beats Music and Beats Electronics in 2014. In 2015, it closed Beats Music and launched Apple Music, with the app installed on all its devices. In doing so, it went into direct competition with Spotify, the leading streaming service. At the time, the 30 percent surcharge boosted Spotify's monthly subscription to the App Store to $12.99, three dollars more than Apple Music's subscription. Spotify was so infuriated, that it sent emails to subscribers with a link to turn off auto-renew for their App Store Spotify subscription and a second link to sign up directly for $9.99.[4]

During the next four years, Spotify accused Apple of making bad-faith changes to App Store policy to impede Spotify's appeal to Apple users. The Swedish music company was unable to sync its product to the Apple Watch, the HomePod, or have Siri—Apple's digital personal assistant—communicate with the Spotify app. Finally, after repeated rejections to its app, Spotify executives appealed to the European Commission in 2019 to level the Apple playing field. The EU issued a Statement of Objections to Apple two years later, "taking a preliminary view that Apple has abused its dominant position."[5]

Similarly, in 2021's closely watched Epic Games vs. Apple lawsuit over the 30 percent surcharge on purchases of the hit game *Fortnite*, California judge Yvonne Gonzalez Rogers ruled that Apple was "permanently restrained and enjoined from prohibiting" mechanisms—buttons, links, calls to action—that provided alternatives to an Apple-only purchase policy. While opting not to call Apple a "monopolist," Rogers wrote: "Nonetheless, the trial did show that Apple is engaging in anti-competitive conduct under California's competition laws."[6]

A SLICE OF SILENCE

I want to talk about Apple and free speech.

Not surprisingly, once again Apple tries to have it both ways with its App Store rules. Here's a passage I found on the App Store Review Guidelines:

We strongly support all points of view being represented on the App Store, as long as the apps are respectful to users with differing opinions and the quality of the app experience is great. **We will reject apps for any content or behavior that we believe is over the line. What line, you ask? Well, as a Supreme Court Justice once said, "I'll know it when I see it." And we think that you will also know it when you cross it.**[7]

I bolded the part that I find particularly disturbing. The most alarming thing is the statement that Apple is the unilateral judge and jury, the arbiter of "inappropriate content." With this stance, Apple marches out on to the slippery slope of censorship. Instead of specifying what "over the line" is, or where it thinks that line is, Apple is essentially saying, "Trust us." And, in a stunning act of audacity—or arrogance—the company then compares itself and its wisdom to the Supreme Court and legendary jurist Potter Stewart.

When it comes to specific rules and policing content, here is Apple's opening salvo:

1.1 Objectionable Content

Apps should not include content that is offensive, insensitive, upsetting, intended to disgust, in exceptionally poor taste, or just plain creepy.

That is about as wide a net of arbitrary rules as any repressive force could wish for. Who decides what is "offensive, insensitive, upsetting"? One person's joke is another person's slight. For Big Tech investors, this entire book is probably "upsetting." According to these rules, op-eds in the *Wall Street Journal* app might fit the bill. A standup comedian on Netflix might get the streaming company's app banned. It's absurd because how content is viewed is always in the eye of the beholder. And if the ultimate beholder is Apple, well, watch out.

Especially if the mega-corporation decides you are a competitor.

Apple's dubious guidelines and wisdom came crashing into the public and political eye on January 9, 2021, when it banned the free-speech app Parler, a favorite app of conservatives since it launched in 2018, from the App Store.

The Apple ban of the Twitter-like app came just three days after the riots inside the Capitol building. Immediately following the uprising, media reports surfaced criticizing the free-speech app for allowing posts that encouraged violence and rallied users to form militia groups and take action against politicians.

"Parler has not upheld its commitment to moderate and remove harmful or dangerous content encouraging violence and illegal activity and is not in compliance with the App Store Review Guidelines."

You could argue that Netflix comedy and crime shows, or many violent video games, do the same.

As it happened, Apple was not the only Big Tech player to target the Trump-loving app. Google also bounced the Parler app for the same reasons. And then, one day after Apple's verdict, Amazon brought the hammer down by refusing to host the site on Amazon Web Services (AWS). In the space of four days, a growing social media site of the American right was shuttered. Canceled. Prevented from competing. Silenced.

Now, I don't agree with the statements on Parler that espoused white nationalist views or other forms of bigotry. I don't like them. I think bigots and racists have no place in my personal vision of America and the Republican Party. But I am not the arbiter of speech, and I don't think that Apple, Google, Facebook, and Amazon are either. If someone has a strong opinion about a politician, they should be allowed to express it. If someone thinks the government has misstepped, they should be allowed to say so. If someone is worried about the integrity of an election—if there was tampering or if Big Tech created informational biases—please, say so.

There is no doubt that Facebook and Twitter were also used by participants in the January 6 Capitol riot, and law enforcement investigators subsequently documented this usage. At no point, as far as I know, did anyone consider shutting down those social media sites. Nor did Google, which indexes, and links to hate speech and racist organizations, consider shutting itself down.

Instead, in keeping with its liberal-loving ideology, Big Tech, evidently upset at the fury of Trump supporters, decided to target Parler. The power and speed with which Parler was silenced is an alarming lesson in the power and speed Big Tech now wields. They control the rules of distribution, they control the infrastructure that allows for distribution, and they control the pipelines to continue service. Those are three choke points to stop speech—any speech—it deems "offensive."

Like many, I was shocked by Big Tech's sudden move to shutter Parler. According to media reports, Apple gave Parler one day to fix its app before yanking it on January 9. Similarly, AWS offers a one-day grace period, before shutting Parler's service on January 10.

Mike Lee, the top Republican on the Senate antitrust panel, shared my concerns about targeting Parler. In early April we sent a letter to Jeff Bezos, Tim Cook, and Sundar Pichai asking for detailed background on the decision-making process for shutting down Parler. In case these CEOs wondered why we were asking, we spelled out the bizarreness of their actions:

In just three days, Apple and Google effectively cut off Parler's primary distribution channel, and Amazon cut off Parler's access to critical computing services, leaving the company completely unable to serve its 15 million users. These actions were against a company that is not alleged to have violated any law. In fact, information provided by Parler to the House Oversight Committee revealed that Parler was assisting law enforcement even in advance of January 6.[8]

Three weeks later, Apple responded saying they would reinstate the app, noting the company was making changes to its content moderation practices.

The censorship of Parler—the unpreferencing—was a high-profile free speech assault, executed in plain sight. Democrat-loving Big Tech used the January 6 Capitol attack as a cover to repress conservative speech. But it turns out there are plenty of other disturbing acts that seek to silence conservative thought that challenges the liberal status quo. And in upcoming pages, I'll document more instances of Big Tech's overt biased behavior designed to limit and restrict the marketplace of ideas.

OPENING UP THE APP MARKETPLACE

Senator Lee and I both support the Open App Markets Act, a bipartisan bill I introduced in the House August 13, 2021, with Congressman Hank Johnson from Georgia. Senator Marsha Blackburn from Tennessee is the lead Republican sponsor of the companion bill in the Senate. These bills prohibit companies that own app platforms with more than 50 million U.S. users from forcing developers to use the in-app payment system owned or controlled by the company. The platform can't demand prices be equal to or more favorable on its app store than another app store, and it can't punish developers for using an in-app payment system or offering different pricing on another app store.

The bill also aims to prevent companies from using platform harvested data on third-party apps to build their own competitive apps, and it cracks down on self-preferencing in search results and ensures preinstalled apps can be uninstalled with minimal effort.

I am confident this bill—assuming Big Tech lobbyists don't succeed in their lobbying campaigns against it and the FTC and Justice departments enforce it—will help ambitious innovators at Tile, Popsockets, Basecamp, and Sonos, as well as behemoths like Spotify and Epic, compete fairly in the marketplace. The innovation they bring must recognize the companies' property rights and risks, not just a monopoly platform owner.

CIRCLING BACK

Self-preferencing on the App store takes many forms and inflicts a wide swath of damage. Apple can use it to shut down competition, as it did with Tile, which in turn shutters innovation. Instead of rewarding a company that invented a new product and was instrumental in creating a new market, the innovator must stop innovating and fight for survival. This impedes the flow of new ideas.

Having a monopoly position in the App Store and closing that market can stop the sharing of new ideas. As we saw with the shuttering of Parler, 15 million Americans can lose their preferred social platform because Apple and Google believe they should have the power over free speech.

Or as I like to describe it: bury the product, bury its view-point. This is the unspoken tactic of Big Tech as they seek to control the markets. What better way than to deny competition, free speech, opposing viewpoints, and logical, substantive criticism.

CHAPTER 7

Insta-Grab

MERGERS AND ACQUISITIONS (AND DECEPTIONS) AT FACEBOOK

I'm shocked, shocked to find that gambling
is going on in here!

CAPTAIN RENAULT IN *CASABLANCA*

It was about 20 minutes before midnight on Monday, February 27, 2012, and Mark Zuckerberg wasn't thinking about sleep. He was thinking about rival social network start-ups like Instagram and protecting his company—the one he'd founded in 2004 after working with a trio of Harvard classmates on a social network called harvardconnection.com. Zuckerberg—who once messaged a friend, "You can be unethical and still be legal that's the way I live my life"[1]—had famously delayed working on that collegiate project so he could focus on his own rival launch: *thefacebook.com*.

Now, Zuckerberg put the finishing touches on an email to Facebook's then-CFO David Ebersman. He was three months away from the largest IPO in history, and he was focused on stomping out competitors. "One business question I've been thinking about recently," he wrote," is how much we should be willing to pay to acquire mobile app companies like Instagram and Path that are building networks that are competitive with our own," he wrote. "The businesses are nascent, but the networks are established, the brands are already meaningful and if they grow to a large scale, they could be very disruptive to us. These entrepreneurs don't want to sell (largely inspired by our success), but at a high enough price—like $500m or $1b—they'd have to consider."[2]

He ended the email by asking for Ebersman's thoughts.

At 9:27 the following morning, Ebersman responded with a series of questions about what Zuckerberg was trying to accomplish: "1) Neutralize a potential competitor? . . . 2) acquire talent? . . . 3) integrate their products with ours in order to improve our service? . . . 4) other?"

Mark Zuckerberg wrote back 40 minutes later: "It's a combination of (1) and (3)."

Then something—perhaps a pang of conscience or maybe a tip-off that he'd just sent two emails that expressly talk about targeting a business competitor—made Zuckerberg write a follow-up email: "I didn't mean to imply that we'd be buying them to prevent them from competing with us in any way. "

Two months later, in early April, competition from Instagram was still weighing heavily on Zuckerberg's mind. Twitter approached Instagram with a $500 million acquisition offer, a seemingly remarkable offer for a company with 13 employees. On April 5, Zuckerberg fired off a series of short emails, including one that read, "I just need to decide if we're going to buy Instagram."

"Gaming is shifting from us to mobile platforms," he wrote minutes later. "It's causing all this negative momentum in a bunch of ways around gamer overall user engagement, ad spend from gamers, overall revenue, etc."[3]

In another missive he summed up: "Instagram can hurt us meaningfully without becoming a huge business though. For the others, if they become big, we'll just regret not doing them."

That potential hurt loomed large. Facebook doubled Twitter's offer and Zuckerberg wooed Instagram's founders with a pledge that they'd continue to operate the company. It would be affiliated with Facebook but operate as a separate company. Within a few years, however, Instagram founders resigned when Zuckerberg forced unwanted changes on their platform.

The Federal Trade Commission essentially waved Facebook's purchase of a rival social network through; its members voted 5–0 to table any complaint.[4] Two years later, Facebook purchased WhatsApp, a leading mobile messaging app.[5] In early 2022, Meta, the recently minted umbrella corporation that owns Facebook, Instagram, WhatsApp, and Messenger, reported 3.64 billion people were using at least one of the company's core products. That power is daunting when you consider three things:

First, these products are, at first blush, terrific. They can unite families and friends, bring people together for charitable causes, and function as a kind of virtual town square. This makes them incredibly seductive.

Second, having won users over, these products use their seductive power to monitor and collect vast troves of user data: what you like; what your read, listen to, and watch; where you live; where you go; what you buy; who your friends are. And they use this data to influence your behavior by selling it for targeted advertising. It is, in a word, manipulative, which is another word for controlling. A monopoly that controls the data that shapes us is possibly the most dangerous monopoly of all. It is self-perpetuating in a sinister fashion. It can gradually bombard a moderate user with more calculatedly liberal—or conservative—content to influence their opinions or, yes, votes.

Third, when a monopoly is headed by someone who has admitted to being unethical, who has an email trail that openly discusses embracing anticompetitive strategies, who has engaged in buy-and-switch practices regarding the user data policies, and who approves of censoring posts when the government tells him to—despite insisting he's a free-speech stalwart—we have a real problem. And nothing gets more to the heart of this sinister problem, for me, as Instagram's amoral, destructive role in steering teenage girls toward depression and suicide.

UNSPEAKABLE HEAD GAMES

The seductive nature of Instagram—sharing and highlighting photographs, which are then distributed according to the company's own algorithmic rules—is dangerous. And Instagram knows it.

Starting in 2019, Facebook has spent three years studying how Instagram's photo-sharing affects millions of teen users. Time and again, according to the *Wall Street Journal*, which uncovered the study, "researchers found that Instagram is harmful for a sizable percentage of them, most notably teenage girls."[6]

The numbers were staggering. Here's the content from slides researchers presented:

- "Thirty-two percent of teen girls said that when they felt bad about their bodies, Instagram made them feel worse. Comparisons on Instagram can change how young women view and describe themselves."
- "We make body image issues worse for one in three teen girls."
- "Teens blame Instagram for increases in the rate of anxiety and depression." "This reaction was unprompted and consistent across all groups."
- 13 percent of British teens and 6 percent of American teens who reported suicidal thoughts traced the desire to kill themselves to Instagram.

As a parent, I can't believe anyone would let a child use a product that's this dangerous.

The conclusions of Facebook researchers from their "teen mental health deep dive" determined that Instagram—as opposed to social media itself—was a breeding ground for negative thoughts, as users may compare themselves to the attractiveness, wealth, and success of others.

"Social comparison is worse on Instagram," researchers concluded, noting that the app highlights users' bodies and lifestyles. TikTok, a rival app, focuses on short-video performances. Snapchat, another photo- and video-sharing rival, lets users deploy filters that "keep the focus on the face."

Researchers also found that the way the site is programmed can also be harmful, especially pages that deploy an algorithm to serve users photos and videos, surfacing potentially harmful images.

None of this should be a surprise, least of all to Mark Zuckerberg, who got in trouble at Harvard for launching a site called FaceMash that let users rate the looks of college students. Anyone who has ever been a teenager knows there is social pressure to fit in, look good, be fit, or have the "right" clothes, hair, and material possessions. Is it surprising that apps filled with glamorous, sexy, luxurious images—no doubt many airbrushed or the result of a thousand discarded photos—might be a negative influence on our youth? If social comparison is real, and I think we can all agree it is, then it's easy to see how Instagram's internal programs can steer users toward eating disorders, depression, and worse.

Or as the researchers at Facebook concluded: "Aspects of Instagram exacerbate each other to create a perfect storm."

Despite that conclusion, Zuckerberg and his minions have done little to nothing to address these findings—even though researchers have suggested cutting back on celebrity content centered on fashion and beauty. They also suggested adding fun filters to selfies instead of "beautifying" filters that allow users to touch up their faces. Their reasoning? Seeing prettified "selfies in stories made people feel worse."

The only remedial action detailed in the *Journal*'s nearly 4,000-word report involved an Instagram pilot program that allowed users to remove the feature in which users "like" an image. Somehow, despite teens flagging "like" counts as a source of anxiety, researchers claimed removing this option helped nothing, "We didn't observe movements in overall well-being measures."

And so, this cancerous menace persists.

Facebook and Instagram have internal data saying their app poses a danger to a vulnerable segment of our population. It doesn't issue warnings to parents. Or to the population at risk. It doesn't change the algorithm, as far as we know. It doesn't share its data with mental health experts. It doesn't post warnings about users taking breaks. It doesn't even admit the app might be a problem.

This irresponsible, amoral behavior starts at the very top, with Zuckerberg and Instagram honcho Adam Mosseri who continue to play disingenuous word games when the mental health of young women is broached. Mosseri was actually quoted saying the app's negative impact on teens is "quite small." Tell that to the girls who spend their lives wrestling with anorexia or bulimia. Or the parents of a suicide victim for whom social media contributed to a tragic downward spiral. As for his boss, well, Zuckerberg came to Congress and was asked if he recognized any connection between children's mental health and social media platforms.

The CEO, who presumably knew something about his own internal research team, kept a straight face: "I don't think that the research is conclusive on that."[7]

With that answer, it seems quite clear that Mark Zuckerberg would prefer to gather more user data—to sell ads and make money—than fix Instagram and stop driving troubled teens to self-harm.

ACQUISITIONS A-GO-GO

Facebook's acquisition of Instagram and WhatsApp weren't made in a vacuum.

Big Tech is not a misnomer. The companies that dominate our society have been on buying sprees for more than two decades now. Facebook acquired 28 companies related to its original business and 77 companies that opened new business sectors. It is also notable that the biggest, most outrageous of Facebook's acquisitions occurred during the Obama administration. I'm not sure whether to describe that as a fitting or tragic coincidence. And maybe it's no coincidence at all. Earlier, I described Obama as the first Facebook president. His embrace of digital culture and digital supremacy has caused significant damage to the marketplace of ideas. As I noted earlier, his choice of a Google executive to run the Patent Office has destroyed innovation instead of protecting it.

As of 2021, Google has made 81 acquisitions related to its original business and 187 acquisitions to help it push into new sectors. In the last chapter, we discussed the 2008 purchase of DoubleClick, which has given the company a stranglehold on ad bidding technology on the web. That same year, Google bought YouTube for $1.6 billion. Today, YouTube makes that much money for Google every three weeks—using that same proprietary ad technology.

While DoubleClick and YouTube were major purchases, Google's most important buy might have been the purchase of Android, a phone software company it bought for a mere $50 million. It is now the dominant mobile operating system in the entire world, with applications that touch everything from phones to watches to computers to cars—and potentially any digital object you can imagine.

Amazon has made 40 acquisitions related to its original e-commerce business and 71 acquisitions that enter new sectors, like its purchases of the Internet Movie Data Base, Audible, and Whole Foods.[8]

Many lower-profile acquisitions have built the retailer's most profitable arm, Amazon Web Services (AWS). Launched in 2006, AWS offers cloud computing infrastructure to any enterprise that needs it. From 2012 to 2020, Amazon acquired 13 tech computing and database companies to support AWS. Today it is the dominant web hosting company in America. The sector now represents the bulk of its profits.

Apple has scooped up 27 companies related to desktop computing and 96 acquisitions related to entering new sectors. In the consumer electronics sector, Apple's largest acquisition in the last decade was its $3 billion purchase of Beats Electronics, the headphone maker founded by rapper and producer Dr Dre. Beats provided the backbone for two major product launches: AirPods, the hugely popular wireless listening devices, and Apple Music, which used Beats streaming service as a foothold to start its battle with music subscription service leader Spotify.

Each of the four monopoly platforms has used the merger and acquisition process to secure its monopoly positions by denying emerging technologies to its competitors. There is little doubt Jeff Bezos, Mark Zuckerberg, Apple CEO Tim Cook, and Google founders (and majority stakeholders) Larry Page and Sergey Brin will defend their respective companies' mergers and acquisitions. They will disavow the M-word. Monopolies?!? Never! They will insist they are just building their corporations, trying to make them more resilient and build shareholder value.

There is also little doubt these leaders know better.

Or as my mama would say, "Their story doesn't hold water."

THE UNFAIR TOWN SQUARE

When one company controls and tracks how nearly four billion people interact on social media, rules matter. They need to be clear and they need to be fair. Facebook failed mightily in this regard at a very important time. And I believe they damaged themselves and our nation.

I'm talking about the rules of engagement—or enforced non-engagement—during the COVID-19 pandemic. Let me be clear: The entire pandemic has been a tragedy. Here in America, millions got sick, at last count, over one million people died. The ensuing lockdowns also had incalculable costs: trillions of dollars were lost. Many businesses got sick and died, too. And the workers who depended on those businesses (and their customers!) also suffered. The number of suicides increased and so did the number of drug overdoses.

And trillions were spent, too—by the government—as it bought debt securities to bolster the stock market and provided aid, care, and even checks. And this added to our already massive national debt.

All of this was happening as we conducted a war with an unknown—or misunderstood enemy, COVID. The questions about the disease were widespread and the answers kept changing. Where did it come from? How did it start? Why did it attack some and not others? How many were at mortal risk? Medical experts, including Dr. Anthony Fauci, flip-flopped on everything from whether masks should be worn to whether it was possible the virus spread was tied to China's Wuhan Institute of Virology.

I'm not making this up. Dr. Fauci did not advise masking in the first months of the pandemic. Then he became the biggest masking cheerleader on the planet. The U.S. surgeon general, Jerome Adams, tweeted that masks "are NOT effective in preventing the general public from catching #Coronavirus." He later reversed himself.

Despite so much uncertainty, the Biden administration insisted it knew better. It lashed out against "misinformation" about the disease, masks, and the vaccine.

And, in a rare instance of Big Tech oversight, it told social media companies what to do.

"We've increased disinformation research and tracking within the Surgeon General's Office. We are flagging problematic posts for Facebook that spread disinformation," Biden spokeswoman Jen Psaki said, adding, "It's important to take faster action against harmful posts . . . and Facebook needs to move more quickly to remove harmful violative posts."[9]

Majority owner Mark Zuckerberg, who constantly pitches himself as a defender of free speech, didn't seem to have a problem with this. His company removed 20 million posts and images on Facebook and Instagram on the grounds they contained COVID-19 misinformation.

Facebook's own help page attempts to clearly state what constitutes misinformation. It gives examples of forbidden thoughts, including one theory that "symptoms of COVID-19 are actually the effect of 5G communication technologies." But the language Facebook uses is frighteningly arbitrary. Especially this:

[W]e remove false information about:

Claims that downplay the severity of COVID-19, such as:

> In the context of discouraging vaccination or questioning the efficacy of vaccines, claims that COVID-19 is no more dangerous to people than the common flu or cold.[10]

When I read this clause, I almost choked.
Facebook was forbidding users to question something?
Really?
Asking questions is not the same as making assertions. Asking questions is part and parcel of the marketplace of ideas. Someone presents a product—a vaccine, say—and people who comprise

the marketplace say, do we need this? Is it good? Is it dangerous? Those are natural questions.

In a prior life I was a prosecutor, not a doctor or a scientist. I have no scientific basis to doubt reports of a COVID vaccine's efficacy. But if I ask a question about that efficacy, I am not spreading misinformation. In fact, it's the opposite. I'm seeking information.

I was alarmed by this silencing of social media users. Not just the Donald Trumps and Rand Pauls who were docked by Facebook and Twitter. But everyone. I wanted more questions asked and answered, not less. In two op-ed pieces for the *Washington Examiner*, I raised questions about the ancillary damage of the pandemic—or our response to the pandemic. Addressing the chronic lockdown with fellow House member Andy Biggs, I wrote: "A third of the country is showing signs of clinical anxiety or depression due to the severity of lockdown measures imposed by governors. More than 40 million workers are now unemployed. It is estimated that half of cancer patients and 80% of brain surgery patients have seen delays in crucial appointments. Schools remain closed, impeding education opportunities and hindering the return to work of parents"[11]

These were all issues that my constituents faced. That all Americans faced and deserved to be able to discuss. But on social media, they risked being treated as pariahs.

SLIPPING ON THAT SLOPE

Together, Meta's companies reach about half the world's population. That is especially remarkable when you consider Facebook and Instagram are not available in China. The Chinese government will not allow its citizens to access Meta apps because Meta—unlike its Big Tech brothers Apple, Amazon, and Google—will not comply with the free speech restrictions that the Chinese Communist Party demands.

Why am I bouncing from COVID-19 to Facebook to China? When Jen Psaki and Biden-appointed surgeon general Vivek Murthy demanded Facebook stop misinformation, they were trying to control speech—just as the CCP does every day in every way throughout China.

I understand there was a health crisis.

I understand there were bad actors selling snake oil cures and ripping people off.

I understand that our elderly population and those with chronic ailments were very vulnerable to COVID.

But in our constitutional republic you cannot stop people from asking questions. A government must seek to answer those questions, not muffle them. It cannot use social media to target politicians or silence politicians because it doesn't like certain opinions. And social media itself must not hold itself out as the jury, the judge, or, worst of all, the executioner of controversial ideas. It must be fair and open.

There's another reason it must be open and fair—neither monopolistic nor government manipulated. Social media sites and apps monitor users, harvesting data. China's totalitarian government operates the same way—collecting all kinds of personal data to assign "citizen scores" to its population. This level of surveillance is Orwellian as citizen scores can be used to punish or reward individuals.

This reality serves as another reminder that while government must protect free speech from Big Tech monopolies, it must also be careful not to dictate the rules of speech at Big Tech companies.

When woke corporate executives working for Big Tech decide what is free speech, it is no longer free.

MONOPOLY AWARENESS AND PREVENTION

For several years, legislators and regulators have bounced around the idea of slowing the scourge of mergers and acquisitions by Big Tech to promote small business growth, investment, innovation, and consumer choice online. The concept is to shift the burden to Big Tech of proving that a merger benefits competition. The Platform Competition and Opportunity Act that was introduced in both houses will, if passed, provide more scrutiny by the regulators and stop Big Tech from binging on competitors by blocking mergers that primarily serve to kill competition and enhance monopoly power.

If Facebook hadn't bought Instagram—a move Mark Zuckerberg's own emails suggest was executed to stop competition—then government would have a more difficult time limiting debate about masks and vaccines. And consumers would have more choices to satisfy their desire for information. Zuckerberg's purchase allowed him to consolidate his role as a social media gatekeeper for half the world.

The fear that start-up companies will not get investors and can't grow—get their product to market—if they are not purchased by Big Tech ignores the reality of the free market. New investment will come from other directions, lured by innovation and new, dazzling consumer choices entering the marketplace, unimpeded, free to be appraised, evaluated, and put to good use.

New companies must be allowed to rise and flourish. Amazon, Apple, Facebook, and Google should know better. They were startups once, too.

CHAPTER 8

Google's Ad-Vantage

COLLUSION, PRIVACY, AND PORTABILITY

The future of advertising is the internet.

BILL GATES

For a pro-life activist like Lila Rose, the founder and president of Live Action, digital advertising is a powerful tool. She can target conservative sites where like-minded readers might donate to the cause. She can plant ads on sites with heavy female readerships, the kind of marketplaces she wants to share her message and her mission.

An activist since the age of 15, Rose founded Live Action, a California-based nonpartisan, nonprofit group, to defend the rights of the preborn and to help the world recognize that abortion is murder. Her site reaches five million people a month, with articles and news about the struggle.

To build on those numbers, and to act on her ultimate mission, stopping abortions and promoting life, Live Action developed an ad campaign promoting Abortion Pill Reversal. The ads were designed to reach women who had taken or were thinking of taking the first of two abortion pills required to terminate a pregnancy.

"Did You Change Your Mind? Do You Regret Your Choice?" read the web ads, which linked to a page on the group's site that explained an alternative—treatment with the hormone progesterone—could save the preborn child.

In September 2021, after the ad had been running for months, Google suddenly banned it. According to Google, the ad suddenly was found to make misleading claims and use restricted terms.

In addition, it pulled the plug on a second campaign promoting "Meet Baby Olivia," a realistic video of a baby growing in the womb, ruling the ad made unreliable claims.

Lila was equal parts furious and heartbroken. She told her quarter of a million followers on Twitter, "At the request of abortion activists, @Google has just BANNED all of @LiveAction's pro-life ads, including those promoting the Abortion Pill Reversal treatment, a resource that has saved 2500 children to date."

Rose blasted the advertising giant for censoring her cause, noting that the liberal news site *Daily Beast* had published an article about the ads prior to the ban. Meanwhile, other pro-life publications reported that Google and Facebook continued to allow pro-choice ads for the abortion pill to run on its platform. "Once again Big Tech is silencing and censoring the pro-life movement," tweeted U.S. Senator Steve Daines (R-Mont.). "This is unacceptable. @Google. must reverse their decision now."

It did—sort of. The company conceded the ad for "Meet Baby Olivia" was banned in error. The Abortion Pill Reversal treatment ad continued to be barred.

Google, in other words, was doing exactly what it and other social media groups did throughout the COVID crisis: it banned free speech that it wasn't comfortable with.

The company may not agree with Abortion Pill Reversal treatment. It can cite members of the medical establishment that say there have been no studies of the treatment so there is no proof the life-saving procedure works. But the counterargument is equally true: *there are no studies showing it doesn't work.* Furthermore, the government hasn't banned the procedure.

Add that up, and liberal, Democrat-donating Google is restricting speech. It's keeping a possible treatment from the marketplace of ideas. For someone like Lila Rose, an activist who wants to celebrate each and every life, denying an ad that is at the core of her life's mission is cruel, crippling, and censorious. It is judgmental and wrong. And it shows just how much power Google has as the gatekeeper of ads.

Google, which owns the second-most trafficked site in the world, YouTube, is also the gatekeeper of videos shown on the web. And it's worth mentioning that site has banned and restricted videos from American conservatives that smack of ideological repression. One video featuring popular radio conservative commentator Dennis Prager discussing the Ten Commandments was tagged by YouTube with a "restricted access" designation, meaning users who have activated YouTube's Restricted Mode will not see the video. "No document in world history so changed the world for the better as did the Ten Commandments," Prager says in the clip. "Western Civilization—the civilization that developed universal human rights, created women's equality, ended slavery, created parliamentary democracy, among other unique achievements—would not have developed without them."

The clip does discuss murder—as in "Thou Shalt Not Kill"—but that is about as PG rated as it gets. Apparently, where Prager is concerned, it's too much.

"We have repeatedly asked Google why our videos are restricted," Prager told Congress. "No explanation is ever given."

This kind of gatekeeping can wound a warrior, silence an idea, and even stop someone from saving a life or praising the Ten Commandments. Every freedom-loving person should be scared by this, regardless of their feelings about abortion or a woman's right to choose. Google can apply the same censorious logic at any time to any ad. This includes politicians. Does Google get to decide which candidates are making "unreliable claims" and pull their campaign ads off the web?

This is just one example of the terrifying potential of Google's monopoly over internet and mobile ads.

But wait—there's more.

PAPER CUTS

Technology that distributes new concepts and products into the marketplace of ideas has come a long way since printing and posting a single sheet bulletin to promote a political movement or protest things like a tea tax. Digital advertising is now the dominant paid marketing tool to influence the marketplace of ideas. While TV advertising is arguably still the most powerful medium—witness the make-or-break sums spent on Super Bowl ads—advertisers shelled out $189 billion on digital advertising in 2021.[1] That number dwarfs the $49.1 billion spent on TV ads[2] and $26 billion in print ads.[3]

In Chapter 5, we looked at how Google's method of indexing newspaper and magazine stories and linking to them has eroded income for many publications and helped fuel the demise of 1,000 weekly newspapers across America. But Google's ad monopoly— its ownership of the system by which publishers list available ad space and advertisers bid on those placements—looms as its most damaging weapon in disempowering the press. Not only can its ad network silence advertisers over self-imposed "standards," it also stands accused of rigging digital ad bids and colluding with Facebook.

Just ask Doug Reynolds.

A former West Virginia state legislator from a well-to-do banking family, Reynolds owns HD Media, which runs several newspapers in the state. In 2018, the company acquired West Virginia's largest paper, *Charleston Gazette-Mail*. The paper was in bankruptcy at the time, but Reynolds, impressed by the paper's 2017 Pulitzer Prize–winning investigative coverage on the ravages of opioid addiction, wanted the paper to endure and thrive.

Reynolds began experimenting. In late 2019, the paper began focusing on building revenue. It added new content, including newsletters and podcasts. "We more than doubled our digital subscriptions. Our readers pay good money to engage with our content digitally," Reynolds said. "We kept reaching more people. But our revenue kept going down."[4]

The more he learned about digital advertising, the more frustrated he became. On January 29, 2021, HD Media filed an antitrust complaint against Google and Facebook, asserting both companies manipulated the digital-advertising market, threatening the survival of the company's papers.

For Reynolds, there was more at stake than his investment. "We invite every other newspaper in America to join this cause," he said. "We are fighting not only for the future of the press but also the preservation of our democracy."[5]

It's easy to get swept up in Reynolds's mission to save democracy—I know I sure do. But let's not forget the future of the press. That's incredibly important. Don't forget: West Virginia has been brutalized by the blight of opioid addiction. The *Gazette-Mail* story that won the Pulitzer Prize shined a much-needed light on that tragic crisis. It took investigative reporters, editors, and photographers to conceive of the report, to research it, to fact-check, to photograph and design. Big Tech firms like Google, Apple, and Facebook don't do that. They just share that work and figure out how to profit from it. Or, as we've also seen, they decide not to share it. So, the importance of an unimpeded news media to America's future cannot be overstated.

But Reynold's found that future in jeopardy: "[T]he newspaper industry over the last 10 years has been making this . . . transfer to digital media, and what we found is, as we've gone into this world, that Google and Facebook make the rules of the game, they control the whole environment."

And controlling the digital ecosystem of ad exchanges means they are in conflict. "They compete against us for advertising dollars," Reynolds says.

And, his suit makes clear, they win those dollars.

"In digital advertising, a single company, Google, simultaneously operates the leading trading venue, as well as the leading intermediaries that buyers and sellers go through to trade," the lawsuit states. "At the same time, Google itself is one of the largest sellers of ad space globally. Google monopolizes advertising markets by engaging in conduct that lawmakers prohibit in other electronic trading markets: Google's ad exchange shares superior trading information and speed with the Google-owned intermediaries, Google steers buy and sell orders to its own exchange and websites (for example, Google Search and YouTube), and Google abuses its access to inside information."[6]

Reynolds's suit also took aim at Facebook, asserting it conspired with Google to "further their worldwide dominance of the digital advertising market in a secret agreement codenamed 'Jedi Blue.'"

ABUSE OF FORCE

This was not the first time Google's covert operation "Jedi Blue" had surfaced in a lawsuit.

Two months earlier, December 16, 2020, Texas attorney general Ken Paxton and nine Republican compatriots from other states filed a federal lawsuit accusing Google of anticompetitive conduct in advertising in violation of the Sherman Act, which outlaws restraint of trade conspiracy and monopolies. The opening of the 130-page suit pointedly noted the company "dropped its famous 'don't be evil' motto. Its business practices reflect that change."[7]

The implication, of course, was that the company had gone over to the dark side; it was, in a sense, evil. The charges that followed were much more specific, detailing many methods in which Google monopolized the sale and placement of web ads. The suit accused Google of seeking "to kill competition and has done so through an array of exclusionary tactics, including an unlawful agreement with Facebook, its largest potential competitive threat, to manipulate advertising auctions."

"Let me put it this way," said Paxton, who spearheaded the suit. "If the free market were a baseball game, Google positioned itself as the pitcher, the batter and the umpire.[8]

But Paxton could have extended the metaphor even further. By attacking competitors, devaluing ad space, and conspiring with Facebook, Google was trying to run the entire league of online advertising and write the rule book, too.

One of the ways Google wanted to rewrite that rule book was by executing Jedi Blue—a secret plan to cut a deal with Facebook and prevent it from giving a boost to a system that Google regarded as an "existential threat," according to internal documents uncovered by investigators.

Google's abuse of monopoly power and its brazen Jedi Blue deal, which I'll detail in a moment, were and are terrifying. Once again, owning and controlling a marketplace was Google's goal. And in this case, the marketplace at risk was vitally important; at their core, ad sales and distribution focus on creating new markets and influencing existing ones. If ads are a form of speech and Google's financial future is focused on controlling the digital ad market, then Google aims to be the gatekeeper of the most potent influencing tool in contemporary society—ads, which artificially affect the marketplace of ideas.

WHAT GOOGLE DID

The buying and selling of web display ads, like financial market trading, involves electronic transactions made at lightning speed. Publishers—websites, mobile apps, and digital displays—offer their ad space inventory on brokerage exchanges. Advertisers then use those exchanges to place bids on the ads.

According to the lawsuit, when Google acquired ad server powerhouse DoubleClick in 2008,[9] it now owned the tools to list display ad inventory—that is, the spaces an ad can be placed on a web page or app or inserted in front of a video—on ad exchanges. As the dominant middleman—Google claims its AdX exchange executes more daily transactions than the NASDAQ and NY Stock Exchange combined—it charged ad buyers a commission fee while also representing the ad space sellers (who also paid a fee). Then it required all transactions to clear its exchange, "where it extracted a third, even larger, fee."

Determined to dominate the market further, Google introduced a new rule: publishers could only route inventory to one exchange at a time. This limited the exposure of available ad space and ultimately devalued it. Instead of reaching buyers on multiple exchanges, an available ad would have to plod through one exchange at a time. "Google demanded that sellers route their ad space to Google's exchange," says the suit, noting the company falsely claimed this would serve the sellers' best interest and maximize revenue."

I want to emphasize this because it demonstrates how Big Tech monopolies adversely impact free speech and threaten our democracy—the heart of why I'm writing this book. *Google devised a rule restricting how a publisher could advertise and monetize available ad space—the very places reserved for potential speech, messaging, and expression.*

Why would Google do this? "Google's real scheme," investigators said, "was to permit its exchange to snatch publishers' best inventory at the expense of publishers' best interests." In other words, publishers' remaining ads were devalued.

In the end, this monopolistic powerplay nearly backfired, inspiring the very innovation monopolies want to thwart. Publishers and developers created a system called header bidding. This blazing-fast software routed ad inventory to multiple neutral exchanges and auctioned it off to the highest bidder nanoseconds before the desired ad space loaded on a given web page.

Header bidding allowed advertisers and web publishers to circumnavigate Google's proprietary ecosystem. By 2017, an estimated 70 percent of web publishers reportedly used the software.

And in March of that year, one critical publisher, Facebook, which maintained the Facebook Audience Network, announced it would join the header bidding revolution.

The shift in the ad market—and Facebook's statement—did not go unnoticed at Google. One ad executive in the company's Publisher Business Group, Chris LaSala, director of New Products & Solutions, wrote a memo outlining 2017 priorities that didn't mince words: "Need to fight off the existential threat posed by header bidding and FAN."[10]

Why was Facebook's ad network a threat? If there is one Big Tech company with more user data than Google, it's Facebook. If the social media giant deployed that data in tandem with header bidding, it might beat Google at its own game.

WAR GAMES

According to the lawsuit, Google countered the "existential threat" of header bidding by running a bait-and-switch operation. First, it lifted the one-exchange rule. This, however, was a ruse, investigators concluded. "Google's program secretly let its own exchange win, even when another exchange submitted a higher bid . . . as one Google employee explained internally, Google deliberately designed the program to avoid competition and the program consequently hurt publishers."

Internal Facebook documents revealed the social media company's header bidding announcement in March 2017 "was part of a planned long-term strategy . . . to draw Google in. Facebook decided to dangle the threat of competition in Google's face and then cut a deal to manipulate the auction." And that appears to be exactly what happened.

Facebook cut a deal with Google bigwigs who dubbed the covert agreement with a codename: "Jedi Blue."

Facebook stopped pursuing header bidding. In return, it was granted "information, speed, and other advantages" in Google-run auctions for publishers' mobile app U.S. advertising inventory, according to the suit. Together both companies would engage in a charade: Facebook and Google faced off head-to-head as bidders in these auctions—but that was for appearances, the attorneys general charge. Their suit states "the parties also agreed up front on quotas for how often Facebook would win publishers' auctions—literally manipulating the auction with minimum spends and quotas for how often Facebook would bid and win. In these auctions, Facebook and Google compete head-to-head as bidders."[11]

Further Sins

If these allegations are true, Google's entire ad system was a giant self-preferencing complex. Google's ad exchange has repeatedly awarded itself and favored partners, such as Facebook, with winning bids, regardless of the competition, according to the lawsuit. The company has also been caught using its core product—search—to self-preference. The title of one *Wall Street Journal* report says it all: "Searching for Video? Google Pushes YouTube Over Rivals."[12]

Reporters found that Google search results consistently featured YouTube video clips over nearly identical clips offered on rival sites—even when a clip on a rival site had overwhelmingly more views and comments than YouTube. In one example, an NBA-issued video clip of basketball star Zion Williamson garnered more than one million views and 900 comments on Facebook. But Google promoted YouTube's version of the video, despite having less than 200,000 plays and less than 400 comments.

"All else being equal, YouTube will be first," a source familiar with Google's inner workings told the *Journal.*

Again, Google denied any hint of wrongdoing. "No preference is given to YouTube or any other video provider in Google search," a spokesperson told the paper.

That's exactly what Google was penalized for in 2017 when European regulators fined the tech giant $2.7 million for abusing its power by prioritizing its own shopping recommendations. Google appealed that decision—and lost in late 2021. It also was whacked with a 50-million euro (about $57 million) fine by a French regulator for murky disclosures regarding how it collected user data and used it for targeted advertising.[13]

What will happen in the complaint against Google for anticompetitive maneuvers to corner the digital ad market? The Sherman Act outlaws "every contract, combination, or conspiracy in restraint of trade," and any "monopolization, attempted monopolization, or conspiracy or combination to monopolize." Texas Attorney General Paxton and his cosponsors clearly believe they have the evidence to prove their case.

DATA DENIAL

While the advertising case is a major weapon of reducing Google's stranglehold on the web, the biggest threat to its monopolistic power is exactly what that power was built on: data. Data on its users. People like you.

Google collects data about you. So does Facebook. So does Amazon. So does Apple. For all their handwringing about privacy protections and data security, these companies use that data to optimize many things. These include the user experience—showing users what they want to see, keeping them engaged—and advertiser impressions and conversion.

When these two focal points are "enhanced," these companies are ultimately optimizing something else: profits. Normally I would cheer for a company making profits. But Big Tech hides behind its "free" services as a reason why antitrust laws shouldn't apply to them. They argue that if their product is free, then anything they do—stifling innovation, acquiring nascent competitors, crushing competition with predatory pricing—can't harm the consumer. In reality, the consumer is harmed by the potential loss of financial benefits. If a consumer's data has such enormous value to these companies, why aren't they paying consumers for that data? The simple answer is that, as monopolies, they don't have to—because they are eliminating regulation, silencing critics, and trouncing competitors who seek the same data for the same purpose.

They are controlling, in effect, the marketplace of data.

But who owns your data? And who should profit from it? These are enormously important questions. The queries you type into a search engine, the sites you visit, the posts you leave on social media, the ads you click on, the purchases you make are all generated by you. Just because Big Tech was shadowing you doesn't give them the right to mine that information and monetize it.

Where's your cut?

The Telecommunications Act of 1996 gave Americans the right to keep their cellphone number when they switch from one phone service provider to another. In effect, as long as you had a working mobile phone and account, you owned your number—a vital piece of data that is specifically tied to you—not the telecom companies. This makes perfect sense.

The trail of vital data about you that you create surfing the web and using apps on your cellphone? You don't own that, but you should. For decades, American companies had to state on their terms and services page that they collected data and what they would use it for. Users who actually read the legal boilerplate usually found they could either opt-out of sharing their data or implicitly consent to its use by simply using the website.

While Google is not selling the specific data it has gathered on you to an advertiser, it is querying that data to serve up targeted ads. "You can say, 'Hey, Google, I want a list of people ages 18–35 who watched the Super Bowl last year,'" says Bennett Cyphers of Electronic Frontier Foundation. "They won't give you that list, but they will let you serve ads to all those people."[14]

So, the people who use Google are providing data that is used to generate targeted ad sales based on the data they have provided. But they have no right to that data.

Since May 25, 2018, the General Data Privacy Regulation (GDPR) has been the law of the land in Europe. Every citizen or resident of the EU's 28 countries has the right to control their personal data. They own it. The GDPR gives users the right to forbid data tracking and to refuse or permit automated profiling. It also allows for the right to data portability, a landmark concept that would allow users to take their Personally Identifying Information and share it—or even sell it—to another site.[15]

None of these laws apply to Americans. Businesses that collect data own the data. While two recent laws out of the West Coast, the California Consumer Privacy Act and the California Privacy Rights Act, give digital users the right to prevent tracking and have their data sold, or even collected, there is no law for portability.

This is a problem. There are other search companies in the world besides Google. Bing, WebCrawler, DuckDuckGo, Neeva, Swisscows, Ecosia. They do not have access to the data about users that Google has. These companies, not to mention many marketing firms, might even pay individuals—creators of data— for access to the records of their searches on Google. But you have no access to it. No rights to it.

Search is important. The faster, more intuitive, and smarter search is, the better. The more information can be accessed. If we want to promote competition, the stranglehold on search-generated data must be ended. Your data should be liberated so that you control who sees it and how it is used.

From where I sit, allowing citizens to control their own data is about as pro-consumer as it gets. Big Tech's data harvesting and targeted advertising has monetary value. The consumers who create that data should get a cut.

A DATA WELFARE STANDARD

To end this chapter, I need to address the precedent in the room.

I'm all for protecting consumers. As far as I'm concerned upholding the Constitution is upholding justice for my fellow citizens. But our nation and its court system must recognize we have entered a brave new digital world where economic power isn't just about a company balance sheet. When a corporation controls the systems through which information flows, when it controls the software that can be placed on phones, when it controls the ads you see or don't see, or the news you see or don't see—then we are talking about a new kind of monopoly power. A digitally optimized monopoly. One that, when I start thinking about all the powerful technology in our future, could be continually optimized by artificial intelligence and machine learning to ensure a monopoly remains.

If this is truly a new kind of potentially monstrous power, then the standards to judge it need to be different. It's not just about protecting prices for consumers. It's about protecting truth, knowledge, and the marketplace of ideas.

It's about freedom.

So, while the consumer welfare standard is currently the litmus test for identifying monopolies, it must be broadened beyond price. And if the courts don't understand this, Congress must act to police of Big Tech monopolies. The members of the House and Senate must recognize the myriad ways a digital monopoly power can damage our society at large. Our lives—from banking, to working, to traveling, to enjoying a movie—are now tracked digitally. This has all kinds of privacy and speech implications. We must pass laws accordingly and prevent the use of technology to reduce our constitutionally guaranteed freedoms. The speed with which the monopolies have coalesced is daunting. We must match that speed to create competition and consumer choice.

Every conservative I talk to in the House and Senate has a slightly different set of solutions for Big Tech abuses, but we all agree that something must be done. Google control and abuse of its position in the digital ad market, is one subject that make my good friend Senator Mike Lee and I equally riled up.

In May of 2022, Senator Lee sent me a draft bill to prevent Big Tech companies from owning more than one part of the digital advertising ecosystem. When he asked me to be the prime sponsor in the House, I jumped at the opportunity. The bill ultimately received the support of other Republicans and Democrats in both chambers. It would prevent tech giants that made more than $20 billion in revenue from operating on all three sides of the ad sales equation: selling, buying, and running the high-speed ad auction exchanges.

I was thrilled the measure was gaining critical support. One press report said this legislation would represent the biggest antitrust change in a generation if passed.

That's exactly the point! I thought as I read the article.

Then I gave thanks for our free press—the one that Google both profits from and harms.

CHAPTER 9

Jungle Predator

HOW AMAZON'S MONOPOLY ABUSES DESTROY COMPETITION, INNOVATION, AND FREE SPEECH

Oil is what this country runs on.
You call it monopoly. I call it enterprise.

JOHN D. ROCKEFELLER

In 2009, Douglas Booms, a senior executive at Amazon, sent an email to three colleagues with the subject line: "Diapers.com—Looked at them ever?"

Booms was flagging a rival company, Quidsi, that had launched a retail site that targeted moms. He dug up some sales numbers that Quidsi "had to expose" during a bid on another business. In four years since 2005, the company had climbed from $2.5 million in sales to a projected $192 million, Booms recounted, adding, "Good growth, no?"

It was a rhetorical question. Those numbers spoke for themselves. Quidsi had proven itself an extremely agile, innovative, and savvy retail start-up. While Amazon was focused on selling everything to everyone as "the world's most customer-centric store," Quidsi founders Marc Lore and Vinit Bharara decided to zoom in on the ultimate shoppers in America—mothers. They launched Diapers.com, placing warehouses near metropolitan areas to reduce the cost of overnight shipping—the balm for the average stressed-out mom. The refined logistics allowed Diapers.com to outperform Amazon on both delivery and pricing. By providing great, low-cost service delivering a product almost every young mother uses, they were forging a customer relationship that would represent a huge lifetime value to the company by building goodwill, and an important database of customers seeking other family-oriented necessities.

Six minutes after Booms hit send on his email, Doug Herrington, then Amazon's vice president of consumables, responded: "They are our biggest competitor in the diaper space.

. . . They keep the pressure on pricing on us. They apparently have lower fulfillment costs than we have. . . . We can approach them through the 'we would be willing to explore a range of relationships' angle."[1]

A few minutes later, Herrington forwarded the email chain to members of his staff. He added a heading—"**Do Not Forward**"—and highlighted the threat posed by Diapers.com: "More evidence these guys are our #1 short term competitor . . . we need to match pricing on these guys no matter the cost."

Amazon did approach Diapers.com and reportedly offered to buy the company. When the offer was rebuffed, Amazon slashed diaper prices by 30 percent.

The diaper price war was on.

Looking back at these emails, however, we can see it wasn't just a price war.

It was about eliminating a competitor—battle to destroy a real threat to Amazon's retailing power.

These emails, which were obtained by our House Antitrust Subcommittee, document out-and-out targeting that moved from a discussion of Quidsi as a possible acquisition target to a predatory pricing strategy designed to crush a rival.

When a publicly traded company is valued at billions of dollars and its rival is a start-up funded by investors, a price war will likely not be a fair fight. Quidsi did not have deep enough pockets to sustain losses on sales of its core product. Amazon did. Another division of the company, Amazon Web Services (AWS), bankrolled the unprofitable retail operation so Amazon could engage in its predatory pricing scheme. More about AWS later.

When Quidsi, continuing to innovate, launched Soap.com, Amazon felt further threatened. Herrington wrote another email detailing drastic action specifically to damage its start-up competitor: "We have already initiated a more aggressive 'plan to win' against diapers.com. . . . To the extent this plan undercuts the core diapers business for diapers.com, it will slow the adoption of soap.com."

Herrington's crew unveiled that plan the same morning Quidsi's founders arrived at an Amazon meeting in Seattle to discuss selling their company. The "Amazon Mom" program debuted, offering customers free membership to Amazon Prime, plus a 30-percent discount on diapers purchased with the store's monthly "subscribe and save" program.

Shocked by the "Amazon Mom" program, the Quidsi team "took what they knew about shipping rates, factored in Procter & Gamble's wholesale prices, and calculated that Amazon was on track to lose $100 million over three months in the diaper category alone."[2]

That projection, however, was way off, according to documents obtained by our subcommittee, which indicated Amazon lost as much as $200 million in a single month.

In the end, Amazon acquired Quidsi in 2011 for $545 million. A month later, Amazon closed the "Amazon Mom" program to new members.[3]

Having achieved their goal of removing a competitive threat, Amazon eventually shut down Diapers.com completely.

LOSS LEADERS AND LOST INNOVATION

The emails I just quoted earlier provide a window into more than just Amazon's tactics fighting for online diaper sales supremacy. They are evidence that the retail behemoth deployed two age-old strategies to exert monopoly control and ensure growth: predatory pricing and competitor acquisition.

By winning market share with its drastically cut prices on diapers—a "loss leader" that brought in customers who were expected to buy items with more profitable margins—Amazon wasn't just hurting Diapers.com's ability to compete; it was threatening its ability to exist.

Proving predatory pricing claims is notoriously hard. The standard set by the Supreme Court's 1993 *Brooke* decision requires establishing that the prices at issue were below its rival's costs and that the rival believed that they would recoup lost profits later. Amazon's diaper offensive seems to meet that standard or come close to it.

Amazon's acquisition of Quidsi is also a clear example of what has become a Big Tech monopoly best practice. Start-ups like Quidsi are vulnerable for mergers and takeovers for two opposing reasons: they don't have enough cash and the market leaders have plenty. When an innovative start-up enters and disrupts a market, it will need cash and investment to sustain itself as more entrenched competitors respond.

Monopolies, by their nature, want to eliminate competition. One of the simplest methods of doing that—of disrupting the disruptor—is to acquire companies that threaten a monopoly's control of a market. And that is precisely what Amazon did with Quidsi. It acquired a company, used it to integrate customers into the Amazon retail ecosystem, and then killed it off.

As I noted last chapter, the consumer-welfare standard—that some monopolistic behavior is okay if it lowers the cost to consumers—has been the guiding tenet in antitrust rulings. This perspective needs to be reevaluated. If, as in the case of Quidsi, the specter of predatory pricing existed before a merger took place, that merger must be stopped and the company that engaged in anticompetitive, destabilizing tactics must be heavily penalized. Imagine if Quidsi's products weren't undersold and the company continued its skyrocketing trajectory. It might have become a rival to Amazon in the diaper market and beyond. Today, consumers would have more choice and lower prices.

Because Amazon's bullying sales tactics worked, there is an argument that the company has disincentivized innovation in the online retail space. How can a start-up compete on pricing or service if Amazon has the cash flow to sustain enormous losses to protect its market share?

Amazon, at least now, does have the cash flow to do just that. In the fourth quarter of 2021, the company reported $137.4 billion in net sales, with an operating income of $3.5 billion. Amazon Web Services had $5.3 billion in operating income on only $18 billion in revenue. In other words, non-AWS businesses lost $1.8 billion during the quarter. That suggests the company's e-commerce and Prime and other divisions are operating at multibillion-dollar losses and bankrolled by AWS.

That underwriting is dangerous. Amazon's market share of e-commerce in the United States is approaching 50 percent. It now has the market power to influence almost any sector it chooses—and stifle competition.

Illegally eliminating rivals is the equivalent to silencing them and stopping consumer choice. Once again, the issue isn't just fairness and profit. It is about the future of American business. When an innovative company like Quidsi must leap over anticompetitive hurdles placed in their way by Big Tech or is shut down, they can no longer innovate.

As a result, the marketplace of ideas suffers.

In an ultra-competitive world where the fourth industrial revolution is currently unfolding and where China has declared its goal of becoming the world's leader in digital technology, fueling an open and free marketplace of ideas is of national importance. Our economy will suffer if ideas are suppressed by powerful companies. And with that, our national security will be threatened.

I'm not the only one who loses sleep about this. "Extraordinary innovation is happening in logistics in the U.S. as big data, automation, and sensors create new possibilities and better results for consumers, businesses and the environment," says venture capitalist Joe Lonsdale. "Competition between new ideas and technologies is transforming our supply chain and attracting tens of billions of dollars in investment. But if a single dominant player is willing to lose billions of dollars a year to snuff out competitive threats, many of the best ideas will lose and innovation will stagnate."[4]

AMAZON: WHEELING, WHEEDLING, AND SELF-DEALING

The Antitrust Subcommittee investigation we kicked off in 2019 had three goals. We wanted to document competition problems in digital markets, examine whether dominant firms engaged in anticompetitive conduct, and assess whether existing antitrust laws, competition policies, and current enforcement levels were adequate to address any issues we might find.

One of our areas of inquiry focused on Amazon Marketplace, the sales channel that Amazon incorporated into its site in 2000 to facilitate the sale of merchandise from third-party sellers. Amazon Marketplace allowed the Seattle store to expand its inventory of available products without actually owning or stocking those items. In return for listing third-party merchandise, Amazon earned a percentage of the sale price of each third-party item. But that wasn't Amazon's only bounty. As a marketplace operator, the company had access to enormous amounts of data on both its shoppers and its third-party sellers. It knew what products customers were searching for, viewing, and buying, and at what prices. But Amazon wasn't just the host of its Marketplace;

it was a competitor, too. Amazon bought items directly from manufacturers and sold them on the site. It also sold and advertised its own private-label products: its Kindle eReader, its Fire tablet, its Glow digital projectors, and its Halo Health fitness tracker. Amazon's dual role as manufacturer and retailer created an inherent conflict of interest. It had the access and ability to abuse its position as the Amazon Marketplace owner to become a better seller within that marketplace. What's more, Amazon managed product search results, determining what items would be displayed first—or last—to every user on its site.

This lack of separation between owning the marketplace and competing against third-party sellers was worth examining—especially in the wake of alarming newspaper reports that claimed Amazon was abusing its powers. Here are just a few of the articles:

- *On* April 23, 2020, the *Wall Street Journal* reported that despite years of claims from Amazon that "when it makes and sells its own products, it doesn't use information it collects from the site's individual third-party sellers," "interviews with more than 20 former employees of Amazon's private-label business and documents reviewed by *The Wall Street Journal* reveal that employees did just that."[5]
- October 14, 2021, *The Markup* reported that although "Amazon told Congress in 2019 that its search results do not take into account whether a product is an Amazon-owned brand," the e-commerce giant "places products from its house brands and products exclusive to the site ahead of those from competitors." It also reported that Amazon regularly listed its own products in organic search results ahead of other products with higher customer ratings and more sales.

The Markup's data-heavy report into Amazon's self-preferencing habits served as confirmation of a blockbuster exposé into Amazon's self-preferencing strategies published by Reuters. That article, "Amazon Copied Products and Rigged Search Results to Promote Its Own Brands," was based on "thousands of pages of internal Amazon documents"—including emails, strategy papers and business plans" and revealed a detailed breakdown of the company's "systematic campaign of creating knockoffs and manipulating search results to boost its own product lines" in India.[6]

A 2016 internal report detailed Amazon's plan to create a brand called Solimo. The document laid out the basic strategy: "use information from [Amazon's India store] to develop products and then leverage [the] "platform to market these products to our customers."

While Amazon associate general counsel Nate Sutton told Congress that his company's search "algorithms are optimized to predict what customers want to buy regardless of the seller," the report obtained by Reuters clearly showed Amazon.in—which is fully owned by Amazon—manipulated search results to give preferential treatment to its own products.

"We used search seeding for newly launched ASINs" (Amazon Standard Identification Numbers) "to ensure that they feature in the first 2 or three ASINs in search results," the report stated, noting the company also used banners known as "search sparkles . . . to specifically promote Solimo products on relevant customer searches from 'All Product Search' and Category search."

In case there's any question about why Amazon would do this, Reuters sources said Amazon's seeding strategy was deployed to boost products—often new items—with low sales ranks. And one former Amazon employee was clear about the impact of these tactics, stating that boosting the rankings of Amazon's own products hurts rival merchants' sales on the platform.

In other words, Amazon's self-preferencing was a documented business practice.

An investigation was fully warranted.

QUESTIONS AND EVASIONS

Over two and half years, our subcommittee heard stunning testimony corroborating these reports from former Amazon employees as well as Amazon Marketplace sellers.

I witnessed Amazon executives and lawyers engage in an infuriating pattern of behavior. Time and again, in writing and testimony, they would assert the company has never used third-party seller data for a competitive advantage and never prioritized its own products in search results. When the committee presented screen shots showing Amazon products placed high in search results, the corporate bigwigs and multiple lawyers stuck to their one-note chorus: deny, deny, deny. In a November 1, 2021, letter to Judiciary Committee chairman Jerry Nadler, Amazon's vice president of public policy, Brian Huseman, dismissed the charges as inaccurate, attributing them to "key misunderstandings and speculation."

Amazon lawyers made a big show of ordering an investigation into its internal business practices. They enumerated all the company's efforts and policies to ensure its Private Brands business unit did not abuse third-party seller information or the power of preferential search results placement. These claims, however, did not disprove the reports in the *Wall Street Journal*, Reuters, or the testimony we had heard from others.

Instead of answering questions, Amazon evaded them or misled our committees—from day one. The associate general counsel for competition, Nate Sutton, began the company's "deny" defense when asked if the company tracked the data of sales items on its site and then created "products that directly compete with those most popular brands that are out there?"

"We don't use individual seller data to directly compete with them [i.e., third-party sellers]," Sutton said, choosing his words carefully. Later, Mr. Sutton reiterated, "We do not use [third-party sellers'] individual data when we're making decisions to launch private brands."

Both statements are ambiguous and misleading. What does "directly compete" mean? Did individual data get aggregated to inform decisions to launch new products?

When Amazon founder Jeff Bezos appeared before us, Representative Kelly Armstrong (R-ND) asked him about the company's internal investigation into the use of third-party data. "Will you commit to informing this committee on the outcome of that investigation, including on the exact circumstances of when Amazon is allowed to view and/or use aggregate data?" my colleague asked.

"Yes," Bezos answered. "Yes, we will do that."

When we asked to see the internal investigation—in the hopes of clarifying these so-called misunderstandings and speculation—nothing was provided. In fact, six weeks later, the Committee received a letter from Amazon's outside counsel that Amazon would not provide any "documents relating to" its most recent internal investigation.

So much for responsiveness from the earth's supposedly most customer-centric company.

NOT SAFE FOR SCREENING—OR READING

Being customer-centric, Amazon is in the entertainment business. It streams movies for its customers to rent, buy, and watch.

Except, of course, when it chooses not to.

The movie business, like the book business, is not just about entertainment. Movies express ideas, they document the world. And so, when Amazon chooses not to offer a movie to its customers, it is, in a very real sense, acting as a censor.

In 2020, a documentary called *What Killed Michael Brown?* was scheduled to debut on Amazon's streaming service. The film, written and narrated by noted conservative Black scholar Shelby Steele and directed by his son, Eli Steele, examined the events and aftermath surrounding the death of Michael Brown, the Black teenager shot dead by a white police officer in Ferguson, Missouri—which led to mass protests against the police.

Steele's documentary critiqued the coverage and language surrounding the tragedy. According to his analysis, mainstream media reported Brown was "executed" and "assassinated." But that narrative is false, says Steele. His film cites forensic evidence, the grand-jury reports and Justice Department investigations clearing the police officer of any wrongdoing. Despite those facts, Steele said, "there are blacks today, right now in Ferguson, as I point out in the film, who still truly believe that Michael Brown was killed out of racial animus."

Two days before the film's scheduled October 16 debut, the *Wall Street Journal* reported that Amazon had placed it under "content review." The following day, the paper reported that Amazon emailed the Steeles, informing them that *What Killed Michael Brown* was "not eligible for publishing" because it "doesn't meet Prime Video's content quality expectations." Amazon went on to say it "will not be accepting resubmission of this title and this decision may not be appealed."[7]

The editorial—catching Amazon in the act of censoring a non-liberal narrative—did the trick. In a matter of days, Amazon Prime reversed itself. Two years later, the film remains on the streaming site with over 1,300 five-star reviews.

One year later, however, during Black History Month, Amazon Prime's slime crew was up to its old tricks. This time it played the suppression games with another conservative Black icon, Supreme Court justice Clarence Thomas, inexplicably deleting *Created Equal: Clarence Thomas in His Own Words*, from its list of documentaries at the beginning of February.

The film, which grew out of 30 hours of interviews with Thomas, debuted on PBS in May 2020. It began streaming on Amazon in October alongside documentaries about Ruth Bader Ginsberg and Thomas critic Anita Hill. But it was pulled after four months. "For a while our film was, briefly, No. 1 in documentaries. And I think it's still No. 25 or 30, so it's been selling," says director Michael Pack, adding that Amazon didn't offer any explanation for the removal."

I'll offer an explanation: Amazon, having perfected the act of eliminating competing products with predatory acts, applied that knowledge to the marketplace of ideas. The Left's narrative about Black History Month focusing on victimhood and struggles was challenged. The life story of Clarence Thomas provides an alternative narrative, with the admirable success of a disadvantaged Black child who overcame poverty to become one of the most celebrated Supreme Court justices. As a conservative, Thomas runs afoul of the Left's narrative. So, liberals at Amazon removed the documentary. We could call this an example of cancel culture. We could also call it eliminating an opponent's narrative.

Amazon doesn't just play Speech God with documentaries. It has also done it with books. A recent example involved the delisting of Ryan Anderson's *When Harry Became Sally: Responding to the Transgender Moment*. After selling the book for three years, Amazon vanquished it, refusing to list the hardcover, paperback, Kindle, or even used copies.

Given that Anderson was the head of the conservative Ethics and Public Policy Center, it was difficult not to wonder if Amazon was yet again targeting a figure from the Right. Responding to a request from Capitol Hill Republicans to explain the removal, Brian Huseman, Amazon's vice president for public policy, wrote: "We have chosen not to sell books that frame LGBTQ+ identity as a mental illness."[8]

Anderson fired back: "Nowhere have I ever said or framed LGBTQ+ identity as a mental illness. The phrase 'mental illness' does occur in the book twice—but not in my own voice: once quoting a 'transwoman' writing in the *New York Times*, and once quoting the current University Distinguished Service Professor of Psychiatry at Johns Hopkins."

I have not read this particular book, so I have no opinion about its content. But I think people should have the choice to read it and reach their own conclusions. Once again, Amazon was acting as a gatekeeper of a product. When the store that controls an estimated 70 percent of all book sales in the United States bans a book, it is affecting that title's impact in the marketplace of ideas.

Thankfully, media pressure paid off after a media report that Amazon had buried a second important documentary about a conservative Black intellectual. *Created Equal* returned to the marketplace of Amazon Prime Video, and with that carved out a potential niche in the marketplace of ideas. But it's hard not to wonder if it was removed because of a secret agenda to block entry to the marketplace of ideas. That, after all, is exactly what censorship, the denial of expression, does. When a giant like Amazon limits the kinds of movies its users can stream, then it isn't just markets that are no longer free. Speech can be curtailed as these giant companies see fit.

LINES IN THE SAND

On March 9, 2022, I cosigned a referral letter with Judiciary Committee chair Jerry Nadler; David N. Cicilline, chair of our Antitrust Subcommittee; and colleagues Pramila Jayapal and Matt Gaetz. It was addressed to Attorney General Merrick Garland, and it alerted him to possible criminal conduct by Amazon and its executives.

After three years of back-and-forth, we were not buying what Amazon was selling: endless denials and misleading statements. We had had enough.

The referral letter was 24 pages long and recounted, with precisely detailed footnotes, Amazon's disturbing business practices and equally disturbing evasions. From our point of view, as legislators charged with writing and upholding the law to serve our nation, Amazon deserved to be penalized for lying to Congress. Here is the key summary from our letter:

> Without producing any evidence to the contrary, Amazon has left standing what appear to be false and misleading statements to the Committee. It has refused to turn over business documents or communications that would either corroborate its claims or correct the record. And it appears to have done so to conceal the truth about its use of third-party sellers' data to advantage its private-label business and its preferencing of private-label products in search results—subjects of the Committee's investigation. As a result, we have no choice but to refer this matter to the Department of Justice to investigate whether Amazon and its executives obstructed Congress in violation of applicable federal law.[9]

This is not a step I took lightly. I've been clear: I like when government steers clear of the marketplace. But when companies conspire to control the marketplace, to squash innovation and avoid competition, then markets are no longer free. And when markets are no longer free, society is no longer free.[10]

What Amazon did to Quidsi was, in my view, anticompetitive. So was its practice of giving preferential placement to its Private Brand products in search results. Meanwhile, by allowing the sale of pirated goods in its marketplace—as jump-rope innovator Molly Metz experienced—it trampled on patent protection and, if the products were made and sold from a foreign country, created a domino effect of lost revenue for rightful inventors and manufacturers.

Our letter to the attorney general targeted Amazon for false statements to Congress. But in my mind, it was also a salvo to Big Tech—a statement that Congress does not believe these new giant companies are immune to the laws of a nation founded in large part to prevent monopoly power.

How can we prevent Big Tech from flexing its muscles and abusing its power to crush competition and innovation?

The two government agencies that investigate and approve mergers, the Antitrust Division of the Department of Justice (DOJ) in conjunction with the Federal Trade Commission (FTC), must rethink their guidelines. In addition to scrutinizing predatory pricing and other anticompetitive chicanery, these agencies must rethink the burden—or lack thereof—placed on mergers involving dominant monopolies.

TOUGHER MEASURES, MORE COMPETITIVE WORLD

While it is too late to undo Amazon's damage to the e-commerce market, we can take action to prevent further abuse. The American

Innovation and Choice Online Act, at the time this book was written, awaits a vote in the House and the Senate. It goes a long way to leveling the playing field, outlawing the unfair preferencing a platform operator's products, services, or lines of business has over those of competitors that are also on the platform. In other words, the self-preferencing we have seen on Amazon, Apple, and Google would be illegal.

The bill also prevents platform operators from excluding products that compete against similar items owned by the platform. And it prevents platforms like Amazon and Apple from using nonpublic data obtained from or generated by a platform to specifically bolster a platform's own products. Finally, to address these monopoly businesses' history of playing hardball, the bill protects users who blow the whistle on these abuses from any form of retaliation.

The bill seeks to be reasonable, allowing companies to take actions if necessary for legal reasons; to protect safety, user privacy, or data security; or to maintain or enhance the "core functionality" of their platform. I'd be lying, though, if I said I didn't worry about them using this broad clause to impede competition in unforeseeable ways.

So, our government watchdogs must act a bit more like attack dogs. They must act to protect a free marketplace of ideas and freedom of speech. Congress must pass the bipartisan legislation that has advanced out of committee. These bills are written to modernize antitrust laws that never contemplated e-commerce and its enormous influence on the economy and speech. Specifically, the law is intended to explicitly prohibit the novel predatory conduct created by these platforms to crush innovation and competition. Monopoly power that allows Big Tech to control markets and the marketplace of ideas presents a grave threat to freedom of speech.

That said, this bill will face an onslaught of criticism and deceptive arguments from the most powerful lobbying forces in America. My colleagues and I need to stay focused on the threat to democracy that is right in front of us. Each time a Big Tech company has a new product, they hone it and weaponize it to control the marketplace of ideas. The product—an operating system, an app, a service, a book, a documentary—is an idea. If they can control who uses that idea and how, they can control that much more of the marketplace. Product placement is inseparable from the placement of ideas.

CHAPTER 10

Double-Talking
Double Standards

HYPOCRISY OF BIG TECH ON FREE SPEECH

———————————————

Doublethink means the power of holding two
contradictory beliefs in one's mind simultaneously,
and accepting both of them.

GEORGE ORWELL'S *1984*

During the brief time I was allotted to question the four Big Tech CEOs testifying at our July 29, 2020, Capitol Hill hearing on anticompetitive practices, I was struck by one exchange in particular: the moment Apple's Tim Cook and Google's Sundar Pichai couldn't bring themselves to provide a simple answer to my simple question.

Maybe my question wasn't as simple as I thought. I mentioned that I was planning to introduce a bill requiring American businesses to certify that their supply chain does not rely on forced labor. Then I said: "While I do not expect you to have intimate knowledge of the legislation, I do want to ask all four of our witnesses a simple yes or no question. Will you certify here today that your company does not use and will never use slave labor to manufacture your products or allow products to be sold on your platform that are manufactured using slave labor?"[1]

"I would love to engage on the legislation with you, Congressman," Tim Cook responded. "But let me be clear, forced labor is abhorrent, and we would not tolerate it in Apple. And so, I would love to get with your office and engage on the legislation."

I had just asked the leader of one of the four biggest companies on the planet for a yes or no answer about using slave labor.

He wouldn't answer.

I couldn't believe it.

"Thank you," I said. "Mr. Pichai?"

"Congressman, I share your concern in this area. I find it abhorrent as well. Happy to engage with your office and discuss this further."

Now I was about to get upset. This was a whole new level of obfuscation. Google and Apple wouldn't disavow slave labor. I decided to give it one more shot.

"I really don't want to even engage with my office half the time," I said. "Will you guys agree that slave labor is not something that you will tolerate in manufacturing your products or in products that are sold on your platforms?"

"I agree, Congressman," Pichai said.

"And Mr. Cook?"

"We wouldn't tolerate it. We would terminate a supplier relationship if it were found."

He still couldn't follow my request! And he would only agree "if it were found."

"I agree," Mark Zuckerberg agreed to agree! Of course, he couldn't just leave it at that, either. The company would have to discover the wrongdoing, a very convenient "if." "We wouldn't tolerate this, and if we found anything like this, we would also terminate any relationship."

"And Mr. Bezos?"

"Yes, I agree completely."

One out of four.

Pathetic.

A SUSPECT COOK

Why was Tim Cook so cagey about answering the most basic of pledges? Who among us would ever want anything to do with slave labor? Honestly, I thought I was tossing them a softball question. One they could easily answer.

But for Tim Cook, the CEO of Apple, slave labor is a tricky issue because the company is so completely enmeshed with China. Tim Cook was measuring his words—not speaking freely—on a number of China-related issues, *including the fact that China is a country that does not allow freedom of speech.* No matter what lip service the Chinese Communist Party pays to individual freedom, the Party and the State are the only things that matter. And so, the CEO of Apple, a company that sources and builds its products in China, subcontracting out to Chinese firms, and that actively sells billions of dollars of its products to China's one-billion-plus population, must choose his words carefully.

Especially when it comes to an issue like slave labor.

That's because Tim Cook undoubtedly knows about the well-documented reports of enforced labor camps in Xinjiang region, where the Chinese Communist Party has imprisoned an estimated one million members of the Muslim Uighur population and even admitted putting "detainees on production lines for their own good" according to the *New York Times.*[2]

Tim Cook and all Apple stockholders should be uneasy. The company's behavior when it comes to China borders has been shameful at times. While its Big Tech brethren Amazon and Google have kowtowed to the CCP, neither can match Apple's betrayal of democratic ideas at the moment of truth.

That moment came in December 2018, when Apple betrayed students and pro-democracy protesters in Hong Kong by bowing to CCP pressure and removing HKmap.live from its App Store. HKmap.live allowed users to upload the position of police trying to silence demonstrators, and its presence in the App Store detonated a firestorm of criticism from Chinese media, which is monitored by and often serves as the mouthpiece for the CCP.

Justifying the removal of an app that enabled citizens to mass and protest without the threat of water cannons and arrests, Tim Cook issued an email to staffers. He claimed the decision to pull the app was made after receiving "credible information" from the authorities "that the app was being used maliciously to target individual officers for violence and to victimize individuals and property where no police are present."

Cook did not share that credible information or which "authorities" provided it. The designers of the app took to Twitter to vehemently deny the app stoked violence and called Apple's move "clearly a political decision to suppress freedom and human rights."

Charles Mok, a pro-democracy lawmaker in Hong Kong, said the app helped everyday citizens avoid police presence. In a letter to Tim Cook, he wrote: "We Hong Kongers will definitely look closely at whether Apple chooses to uphold its commitment to free expression and other basic human rights or become an accomplice for Chinese censorship and oppression."[3]

As far as I know, no media reports surfaced that confirmed that the app posed a danger to authorities. Meanwhile, isn't it curious that Apple, a company that constantly talks about protecting individual security, didn't care about the security of its users in Hong Kong who might have wanted to avoid repressive police actions?

Maybe "curious" is the wrong word. Maybe "hypocritical" is the right one.

In August 2020, Apple released a statement entitled "Our Commitment to Human Rights." The first heading—I am not making this up—was "People Come First."

Here are the opening lines. "Our respect for human rights begins with our commitment to treating everyone with dignity and respect. But it doesn't end there.

"We believe in the power of technology to empower and connect people around the world—and that business can and should be a force for good. Achieving that takes innovation, hard work, and a focus on serving others.

"It also means leading with our values."[4]

It goes on and on with other earnest declarations. Make that *laughable and hypocritical earnest declarations.*

Yes, let's talk about Apple's values, the values of the most highly valued company in the world.

We've already seen how Apple's "values" didn't stop it from betraying the pro-democracy protesters in Hong Kong when it removed the HKmap.live app. Those same values were nowhere to be found in 2021 when two apps, the Olive Tree Bible App and the Quran Majeed, were removed from the App Store in China, according to Apple because Chinese officials believed the apps breached Chinese laws on hosting illegal religious texts.[5]

That's right: Apple was told the Bible and Koran were illegal. So, the company followed orders of what it would term a "complex issue."

But don't hold Apple responsible. Oh, no. Here's how they explain away any responsibility in their Human Rights document:

"We're required to comply with local laws, and at times there are complex issues about which we may disagree with governments and other stakeholders on the right path forward."

That's certainly true. But nobody required Tim Cook and Apple to manufacture products in one of the most repressive nations in the world, a place controlled by a government that has expressly stated that freedom of the press and freedom of religion pose a threat to the CCP.

DOCUMENT 9

If Tim Cook or Apple stockholders have any doubt about the values of the CCP that they, along with so many American companies, willingly enrich, perhaps they could seek out a paper issued in 2013 by the CCP's General Office with the unwieldy title of "Communiqué on the Current State of the Ideological Sphere."

Since it was the ninth paper issued that year, most China scholars know it as Document 9.

Calling the state of the world a "complicated, intense struggle," the paper lays out the CCP's stance concerning "false ideological trends, positions, and activities." What follows is an unabashed attack on concepts like Western constitutional democracy, "universal values," the promotion of civil society, and freedom of the press. According to the CCP here's the big problem with freedom of the press: it can "undermine our country's principle that the media should be infused with the spirit of the Party."

The value of the press, per the CCP—are you listening, Tim Cook?—is that it should be a tool of government control. How is that for a value?

Document 9 is a frightening mission statement. You can search for it online. It explains the relentless flow of human rights abuses in China. Where Tibet doesn't exist, where you can't say Taiwan, where you can't download a Bible or read about the Tiananmen Square massacre. Here's Document 9's final call to action:

> We must reinforce our management of all types and levels of propaganda on the cultural front, perfect and carry out related administrative systems, and allow absolutely no opportunity or outlets for incorrect thinking or viewpoints to spread . . . strengthen guidance of public opinion on the Internet, purify the environment of public opinion on the Internet. Improve and innovate our management strategies and methods to achieve our goals in a legal, scientific, and effective way.

Some will say that Tim Cook, whose company has tens of billions in sales in China annually, is being fiscally prudent. And this begs a central question of this book: Is it in the consumers' best interest to allow a monopoly to control the marketplace and speech? The commercial marketplace and the marketplace of ideas operate in the best interests of consumers when there is the freedom to choose.

This is the line that has surfaced time and again as I researched this book. Story after story cast a shadow over these companies— companies that, let's be honest, are terrific, innovative entities who have made a lot of people rich and improved lives of employees and customers. But for all the good they do, their quest for market control involves self-preferencing, market manipulation, and anticompetitive business practices. Open up the ideological engines of these companies and it's like there's a vacuum sucking out any sense of balance, of propriety, of fairness, of decency. That is the curse of monopoly power. It just wants to feed itself. There are no checks and balances within the market because the market has been locked up.

SEARCHING FOR CONTROL

Apple isn't the only Big Tech company playing fast and loose with the levers of democracy. In the last chapter, we discussed Jedi Blue—a suspected collusion operation with Facebook to stave off a challenge to Google's ad distribution monopoly. Now it's time to talk about the hypocrisy of operation Dragonfly—Google's secret program to launch a censored version of its search engine in China.

That's right: the company that once pledged "don't be evil," cut a deal to bring its best-of-breed search engine to China and block every single thing the CCP finds objectionable. The terms of Dragonfly called for Google's search to sync with all the banned subjects and content excluded by the Great Firewall of China— the government-administered buffer to the Western world.

That meant that information about basic human rights, freedom of the press, democracy, criticism of China policy or China's leaders, any mention of the Tiananmen Square massacre, would not show up in Dragonfly's searches. Information deemed problematic or unacceptable would be suppressed, airbrushed, whited out for users in China.

Even more problematic was the idea that Google, which tracks and monitors data fanatically, would be able to monitor and identify users who searched for terms banned by the CCP. Activists and human rights experts worried that the search engine would operate as a targeting tool for China's Big Brother–like surveillance state.

Dragonfly went beyond the planning stages. According to The Intercept, the liberal site that broke the story, development on the project accelerated after Google's CEO, Sundar Pichai, met with a top Chinese government official in 2017. A cadre of Google coders produced apps with names like "Maotai" and "Longfei" to demonstrate the censored, China-compliant version of Google.

From a business perspective, you can understand why Google might want to do business in China. The CCP is harvesting every scrap of data relating to China's 1.4 billion population. And Google is being shut out of that. If Sundar Pichai is losing sleep over the fact that Tim Cook and Jeff Bezos have more access to Chinese data than he does, he can relax. In the CCP's eyes, Apple and Amazon don't own that data. The CCP does. In 2016, the government passed a law essentially saying that all data created within China's borders was the property of the CCP. Article 37 of the law says all businesses in "critical sectors"—businesses involved in communications, information services, energy, transport, water, financial services, public services, and electronic government services—are required "to store within mainland China data that is gathered or produced by the network operator in the country.

In addition, the law also requires business information and data on Chinese citizens gathered within China to be kept on domestic servers and not transferred abroad without permission."[6]

So that data Apple may collect on clients in Beijing? They don't really own it. The CCP does.

I'm not sure Sundar Pichai should feel better about that fact. He should have been so disturbed; he should never have met with a CCP bigwig in the first place!

Fortunately, the leak about Dragonfly—and the obvious concerns about America's top data company buddying up with the world's number one surveillance state—didn't sit well. A year later, in 2019, Google executive Karan Bhatia delivered some welcome news to the U.S. Senate Judiciary Committee. "We have terminated Project Dragonfly."

It is interesting to know that while Google cooperated with the Chinese Communist Party, it refused to renew a contract with the U.S. Department of Defense to work on the Pentagon's Maven project, which uses artificial intelligence to interpret video images. Apparently, pacifist programmers at the search company were worried their work might be used for advanced weapons or drone strikes, and so 4,000 workers signed petitions demanding "a clear policy stating that neither Google nor its contractors will ever build warfare technology."

Judging by the silence of programmers over Dragonfly and the upheaval over Maven, it seems like employees there have things backward. "Chinese Censorship good, U.S. military security bad."

But maybe that attitude makes sense. China is a surveillance state that gathers data on its citizens to maintain control over them. Google does something very similar—it gathers massive amounts of data on citizens to sell them a click to a car, a candidate, or whatever else someone is hawking.

The hypocrisy hit a high point when the company refused to enter a request for a proposal solicitation for the U.S. Department of Defense's JEDI cloud computing system. Again, though, after a backlash, Google's head of cloud computing recently claimed that if invited to bid on the newly renamed JWCC contract, "we will absolutely bid," and if it wins that bid, "we will proudly work with the DoD" because "we believe Google Cloud should seek to serve the government where it is capable of doing so and the work aligns with our AI principles and company values."[7]

Values? *Company values*? What are those?

Like Apple, Google has a Human Rights document. Here's the beginning of the second paragraph. "In everything we do, including launching new products and expanding our operations around the globe, we are guided by internationally recognized human rights standards. We are committed to respecting the rights enshrined in the Universal Declaration of Human Rights and its implementing treaties. . . ."[8]

Okay! Let's pause right there and cut to that Universal Declaration, which riffs of the original UN charter stating everyone deserves freedom of speech, of religion, from fear, and from want. Article 18 states, "Everyone has the right to freedom of thought, conscience, and religion"; Article 19 says: "Everyone has the right to freedom of opinion and expression; this right includes freedom to hold opinions without interference and to seek, receive and impart information and ideas through any media and regardless of frontiers."[9]

How could the company say it believes in these ideas and then go build a restrictive search engine for China? It didn't hold those values until *The Intercept* blew the whistle and the company looked ridiculously bad. It valued the DoD until its engineers all threatened to quit. It values openness and making all data free to crawl, index, and share, but if individuals want to take the data

Google has collected and give it to another search engine or ad exchange? Er, that's a little problematic. It values completely open systems and zero patents—until a rival ad exchange threatens it.

Let's be honest: the most important *value* of Apple and Google and all these companies is the *value of their stock price.* The driver here is making sure that there isn't a competitor that challenges the price structure for their products. This means it is more important than ever for these companies to control the marketplace of ideas—monopolizing a market means you can control the ideas that go into the market. And therefore, you can control the perception of your own value. In this way, monopoly power can snowball to ensure even greater power.

Interestingly, there's only one place this doesn't work. China. Because the CCP controls the marketplace of ideas there. If the CCP doesn't like an idea or a company, it's gone. Banned. Look at Jack Ma, the founder of Alibaba, the company with one of the largest IPOs in history. He fell out of favor with the CCP and was jettisoned in a flash.

Look at the Tang Energy Company, a Texas firm that cofounded the turbine propeller company HT Blade with the Aviation Industry Corporation of China. Months before a planned $2 billion IPO, HT Blade's assets were unilaterally distributed to rival Chinese companies by the State-Owned Assets Supervision and Administration Council. Tang was left with nothing.[10]

U.S. companies—especially Apple and Amazon—are so desperate to sell to China's vast market of consumers and profit off its low-cost labor, that any "values" they might have—like protecting intellectual property, upholding free speech, not using forced labor—are nothing but hot air when it comes to China. Like Jack Ma, they can be jettisoned.

Tim Cook knows all this. Jeff Bezos knows all this. Sundar Pichai knows all this. And so does Mark Zuckerberg. Zuckerberg's companies are not active in China. This is one point in Facebook's favor. But Zuckerberg is as inconsistent as the rest of his monopolist buddies. "In general, I don't think it's right for a private company to censor politicians or the news in a democracy," he said in an October 17, 2019, speech at Georgetown University, that noted Martin Luther King Jr. "was unconstitutionally jailed for protesting peacefully." "In the end, all of these decisions were wrong. Pulling back on free expression wasn't the answer and, in fact, it often ended up hurting the minority views we seek to protect."

Once again, a Big Tech CEO was publicly promoting free speech. But as this book has documented—where are those values when Facebook censors speech? When it bans posts and politicians, including the former president of the United States? When it refuses to change its photo features to prevent teenage suicide?

It's enough to make you think their greatest value isn't free speech.

It's two ancillary liberties they love: freedom of hypocrisy and from accountability.

CHAPTER 11

The Swamp

ALL OF THE ABOVE

We draw the line against misconduct.
Not against wealth.

THEODORE ROOSEVELT

T wenty-nine hours.

That's how long the markup session took for the House Judiciary Committee to amend and pass six bipartisan bills that aim to rein in the monopoly power of Big Tech.

The media described this as a marathon session. That was an understatement. We started at 10 a.m. on June 23 of 2021 and stopped at 8 a.m. the following day. Then we resumed the session later that afternoon.

You might wonder why, if these were bipartisan bills, did the session take 29 hours? After all, most sessions take a few hours at most.

That's a great question. There are couple of possible short answers.

One is that there was an unusual amount of procedural obstruction, with countermeasures and proposed amendments, all of which wasted a great deal of time. Committees only go into a markup session when members think they have the votes to move something forward. If a member has a serious proposal they want to add to the bill, they will float it to everyone on the committee beforehand so we can perform our due diligence. A last-ditch change might be a fantastic idea, but responsible members will be hesitant to vote for something they haven't fully vetted.

Another answer is that Big Tech was trying its darndest to derail the proposed measures. As I noted in an earlier chapter, in 2021, Meta, which owns Facebook, spent $20,070,000 on lobbying, Amazon plunked down $19,320,000—which makes them among the top 10 influence buyers on Capitol Hill. Meanwhile, Alphabet,

which owns Google and YouTube, and Apple spent big, too, paying out $11,770,000 and $6,500,000, respectively.[1] Lobbyists can
contribute $5,000 to each member of Congress during an election
cycle, plus another $10,000 to election committees.

A third answer: members of Congress from California.

A fourth answer is all of the above.

Let's focus on the California delegation. Take, for example,
Democratic Representative Zoe Lofgren.

I like Representative Lofgren. She is a very nice colleague. The
congresswoman, however, is like a lot of her fellow California legislators: she is extremely protective of Big Tech.

That's because, as she proudly proclaims on her web page, she
"represents the 19th District of California, based in the 'Capital of
Silicon Valley,' San Jose, and the Santa Clara Valley." That means
she represents ground-zero for Big Tech; her district straddles the
voting districts of Apple, Facebook, and Google. Those companies pump hundreds of millions of dollars into the local economy.
They fuel an explosive real estate market. Their presence is part of
the reason why Silicon Valley provides 10 percent of California's
jobs.

Also, more disturbingly, her daughter serves as an in-house
corporate counsel at Google.

I have no doubt that Representative Lofgren is 100 percent
pure of heart when it comes to her legislative choices. She may
believe they are good for the country and her constituents. But
the optics of her own family's connections don't inspire con-
fidence—especially when you consider her efforts to derail the
Antitrust Subcommittee's work.

"It's problematic," says Jeff Hauser, director of the Revolving
Door Project, which examines money in politics. "It's definitely a
conflict of interest and if I were Lofgren, I'd want to address the
concerns by recusing myself—not taking a leadership role."[2]

But that's exactly what Representative Lofgren *didn't do*: she led the opposition. In the June 23 markup hearing, she voted for almost every poison pill amendment, offered last-minute amendments of her own, and argued in favor of any effort to derail the bills. Since the markup hearing she's met with various Democratic caucuses, praising Big Tech's unbridled power, and using Big Tech's talking points about how these companies are really benevolent monopolies.

BIG TECH, BAD OPTICS—EVERYWHERE

Zoe Lofgren isn't the only Democrat compromised by Big Tech. Nancy Pelosi, the speaker of the House as I write this, is married to tech investor Paul Pelosi, a man who has invested millions of dollars in Apple, Alphabet, and other tech stocks. He has profited greatly as these stocks have climbed in value.

On June 18, 2021, just five days before the Antitrust Subcommittee markup, Paul Pelosi bought 4,000 shares of Alphabet via a call option—in which he promised stocks at a later date at a price of $1,200 a share. A month later, the shares closed just over $2,500, making Pelosi $5.2 million richer—without spending a penny.[3]

As you might expect, Speaker Pelosi's office issued a statement: "The speaker has no involvement or prior knowledge of these transactions." That may be true, and I may be LeBron James's twin brother. The fact remains that Speaker Pelosi has brought hundreds of bills to the House floor for a vote in the months following the antitrust markup. None of the antitrust bills were considered.

It's a safe bet that the Pelosi family were feeling warm and fuzzy about the bet Paul made and Alphabet's timely growth. They profited from that growth. Alphabet's and Google's value increased the Pelosis' value. Basic human nature suggests that on some level the Pelosis like Big Tech, or the money it generates for them. All of which makes Nancy Pelosi look compromised.

Curiously, on June 29, 2021, soon after Paul Pelosi's stock purchase, Democratic House Majority Leader Steny Hoyer of Maryland—the speaker's deputy—said the Judiciary Committee's newly passed bipartisan bills were not ready for a House vote. "Right now, they're not ready for the floor, and I don't want to make a prediction as to when they're going to be ready."

This is how conspiracy theories start.

Unfortunately, Nancy Pelosi can deflect criticism easily. Why? Because she can say that America is filled with people who invest in Alphabet. But that's not her strongest defense. Oh, no. Her strongest defense is this: "Hey, I'm not so bad. Look at the guy running the Senate!"

And she would be right, too. The daughter of Democrat Senate Leader Chuck Schumer is—are you sitting down?—a registered lobbyist for Amazon.[4] That's right. When the most powerful man in the Senate sits down for Thanksgiving dinner—or any family dinner—Jeff Bezos's lobbyist has a seat at the table.

Does that sound right to you?

It doesn't to me, either.

Oh, and Schumer's other daughter works for Facebook.

I wish I were making these things up. But I'm not.

Like Pelosi, Chuck Schumer is sitting on more than half-a-dozen antitrust bills. Bipartisan antitrust bills. Here's the thing: if either leader moves these bills forward, it will put enormous pressure on the other leader to follow suit. There may be several reasons for the lack of action, but two reasons come to mind immediately: self-interest and enormous pressure from Big Tech. As for the most powerful Democratic Party member, President Biden—who declared in his 2021 State of the Union address that "we must hold social media platforms accountable for the national experiment they're conducting on our children for profit"—he can and must pressure his party's congressional leadership.

When I hear about these connections, I start to think that Big Tech companies aren't the only ones who self-preference. Our politicians don't seem immune, either. This reflects poorly on all who serve.

Sometimes I think that if I were a swamp, I'd be insulted by all these comparisons to Washington.

UNWINDING THE COMPLEX WEB

I want to pause and make three things very clear:

1. I am not anti-profit.
2. I am not anti-business.
3. *I am pro-competition.*

I believe I've been very clear about all this in these pages. But because I'm focusing on Washington—and I am part of Washington—I don't want anyone to be confused here. So let me say this again:

I believe that when the companies controlling the platforms for free speech and political speech use their power to become arbiters of permissible speech, it is a threat to America and liberty guaranteed in our Constitution. Throughout American history, there has been a delicate balance between the concentration of economic wealth and the dispersion of political power. Amazon, Apple, Facebook, and Google have acquired control over essential infrastructure of our commerce and communications. These monopolies have concentrated wealth in sectors of our economy that result in concentration of power over the dissemination of information and infringe on the free flow of ideas.

I just noted how Big Tech's tentacles extend into Congress. (I don't mean to suggest that Lofgren's and Schumer's daughters aren't fine, talented individuals.) These hires raise obvious questions about Big Tech seeking to gain power and influence. And these companies are betting that most Americans won't know or care about their strategy. That's how monopolies work: nothing gets in their way even when they've done or are doing something clearly wrong.

Now Big Tech has devised more sophisticated and hidden influencing methods, including outsourcing to the most powerful lobbying organization in America.

I'm talking about the U.S. Chamber of Commerce. In 2021, the U.S. Chamber spent $66 million on lobbying. Expect that number to balloon even more. Because the organization is now on the payroll of Big Tech.

The Chamber of Commerce was created at the suggestion of President Howard Taft to represent business interests in Washington. Over the years, it has been advertised as a mainstay of pro-business policies. It endorses Republican and Democratic candidates. The reality is that the Chamber is pro-business in ways that help its board members and other contributing members. It selectively espouses a philosophy of antiregulation and small government (despite, ironically, spending more than anyone to influence that government).

For most of the digital era, the Chamber has not weighed in on tech issues. But with antitrust bills looming, the organization has suddenly "decided" to leap to Big Tech's defense, launching attacks on the Federal Trade Commission.

"Any time we see a big expansion in regulatory power over industry, we're going to get involved," says Suzanne Clark, the Chamber's chief executive officer. "We didn't choose the FTC and antitrust. That battle came to us."

Coincidentally, so did the money. Meta Platforms, Inc.'s chief privacy officer Erin Egan, reportedly a close pal of Clark, recently joined the Chamber's board of directors. The Chamber does not reveal member donations or fees, but Bloomberg reported that Amazon started contributing in 2021. The board Egan sits on is filled with emissaries from America's most powerful corporations.

"The chamber is now just another tool of Google, Facebook and Amazon, and they should be ashamed for allowing themselves to be captured by the most powerful corporations," says Barry Lynn, the director of Open Markets Institute in Washington.

With leaders like Pelosi and Schumer so enmeshed with Big Tech, with Democratic California donors pouring millions of dollars into Washington campaign funds, and with an all-powerful ally like the Chamber of Commerce, it's easy to think that Big Tech has positioned itself to withstand and defeat an antitrust backlash.

But I'll be darned if that's the case.

Fortunately, I'm not alone. I expect Democrats, like my liberal colleague Representative David Cicilline and Senator Amy Klobuchar, to pressure their leaders. And if they don't, I know leading Republicans will also take to the Hill to make a stand against Big Tech hegemony. Indeed, you would think all Democrats would sit up and take notice when they discover that the most liberal members of their party agree with Ted Cruz.

"Big Tech, I believe, poses the single greatest threat to free speech in our country today. And poses the single greatest threat to democracy in our day," the Texas senator says. "Big Tech are the robber barons of the twenty-first century. Any other industry that engaged in the brazen abuse of power we see every day from Big Tech, any other industry with wanton disregard for anyone's rights [that] silenced, shadow banned, demonetized other players, other actors, other speakers—any other industry would face a mountain of lawsuits for their brazen abuse of power and abuse of monopoly power in the market."[5]

The case against Big Tech should be opened and shut.

But it needs to get to the floor of the House and the Senate: the marketplace of legislative ideas.

REMEMBERING OUR HISTORY

On January 17, 1925, President Calvin Coolidge gave an address to the American Society of Newspaper Editors in Washington, D.C., that was the source of his most famous quote: "The business of America is business."[6]

Attributed to a small-government conservative, this line is often taken as a succinct encapsulation of the philosophy of the U.S. government. It makes sense, right? America has grown through its business.

But Coolidge was talking about the people of our nation—not government doing the bidding of corporations. "After all," he said, "the chief business of the American people is business. They are profoundly concerned with producing, buying, selling, investing, and prospering in the world."

The bulk of Coolidge's address wasn't about business at all. It was about freedom of the press and the line between truth and propaganda. Even after one hundred years, it's fascinating to find that many of the ideas he touched on still echo through political discussions about antitrust policy in general and specifically the issues covered in this book. "Wherever the cause of liberty is making its way, one of its highest accomplishments is the guarantee of the freedom of the press," he said in his opening statement. "It has always been realized, sometimes instinctively, oftentimes expressly, that truth and freedom are inseparable."[7]

He talked about people worrying that wealthy publishers might manipulate news in a self-serving manner: "They note that great newspapers are great business enterprises earning large profits and controlled by men of wealth. So, they fear that in such control the press may tend to support the private interests of those who own the papers, rather than the general interest of the whole people."

He noted the importance of separating publishing of information from ownership interests: "Editorial policy and news policy must not be influenced by business consideration; business policies must not be affected by editorial programs."

And he said there is "no justification for interfering with the freedom of the press, because all freedom, though it may sometimes tend toward excesses, bears within it those remedies which will finally effect a cure for its own disorders."

Coolidge was a fan of laissez-faire economics. He believed, as I do, that the free market would take care of itself. As history would have it, he presided over explosive financial growth.

With all this in mind, let's get back to that original misquote. "The business of America is business" is a dangerous precept when it is misconstrued. It implies that business has a life of its own. That business should be beyond the reach of our elected officials when they are acting in the best interests of American consumers, and that business is more important to America than liberty, freedom, and the elimination of tyranny. This is, in my view, fundamentally wrong. The fact is that liberty and freedom promote business, growth, and innovation. Feeding the marketplace of ideas has spurred growth and built business. Not the other way around.

Thinking that business is the only thing that matters—is dangerous. In the era of lobbying and sophisticated influencing

operations; agenda-fueled think tanks; of patronage and placement; and truly, of inconceivable wealth creation, Coolidge's
pithy statement has become a first commandment. But he wasn't
talking about businesses. He was talking about the American
people and their desire to strive and thrive.

The business of America, I remind us all, is spelled out in
the preamble of the Constitution. My favorite passage in all of
America's foundational documents is:

"We the People of the United States, in Order to form a more
perfect Union, establish Justice, ensure domestic Tranquility,
provide for the common defense, promote the general Welfare,
and secure the Blessings of Liberty to ourselves and our Posterity,
do ordain and establish this Constitution for the United States of
America."

This book is about securing the Blessings of Liberty for ourselves and our posterity. And not entrusting them to monopolies
that control the marketplace of ideas.

I believe we can do that. Republican, Democrat, Independent.

Our birthright of liberty is in grave danger. We must work
together, as we have done for nearly 250 years, to form a more
perfect union. If we cede power to corporate monopolies that
control the way we communicate, then the cherished blessing
we profit from every day, freedom—of speech, of religion, of the
press, of ideas—will vanish.

CHAPTER 12

Obligation of Democracy

A CITIZEN'S GUIDE
TO REINING IN BIG TECH

It is not in numbers, but in unity,
that our great strength lies;
yet our present numbers are sufficient
to repel the force of all the world.

THOMAS PAINE

*P*olitico magazine published an article in December 2021 called, "The Congressman Who Doesn't Use Google." It was about a member of the U.S. House who had been staging, or trying to stage, a one-person Capitol Hill boycott of a set of companies most of Washington would find nearly impossible to give up. The reporter detailed how this "quirky member of Congress" didn't use the Google search engine, requested that his staff not shop with Amazon, and "doesn't use Facebook, even to communicate with family."[1]

What wasn't in the story, however, was how the magazine found out about this crusading congressman in the first place. He didn't have a PR team or aide pitching him. He wasn't hijacking press conferences or caught doing backroom deals. No, *Politico* realized there was a story when the congressman was booked to appear online at *Politico*'s Tech Summit: "At a Digital Crossroads." Ten minutes before the scheduled 30-minute talk, the congressman was instructed to log on to Google Meet, which *Politico* was using to conduct the webcast. There was one problem: the congressman didn't have a Google account.

The *Politico* Summit producer coordinating with the congressman was dumbfounded. There was silence on the other end of the phone. The congressman imagined a group of young techies shaking their heads and thinking: *What kind of politician is this guy? Everybody and their great-grandma has a Google account.* After a moment, the congressman repeated himself: "I don't have a Google account."

"Well, that's interesting," the *Politico* producer finally said.

It took 25 minutes for a work-around to be arranged, and the congressman talked for only about 12 minutes of his allotted 30 minutes.

After the event, *Politico* editors learned of the congressman's Google-less reality and thought: *We've got to cover this guy.*

That congressman was me.

I'm biased, of course, but I thought the reporter did a fine job portraying me as a man on a mission—the ranking Republican member on the House Judiciary Committee's Antitrust Subcommittee.

She also exposed just how hard it is to avoid Big Tech, reporting that I used an Apple iPhone because the House of Representatives' IT department gives members a choice of two phones—iPhones and Samsung phones running Google's Android operating system. The article noted that I had to make a choice between two Big Tech giants, and that iPhones have a 99 percent market share of the 10,000 phones in the House.

The reporter didn't mention, however, that there are almost no alternative smartphones that don't depend on either the Android or iOS systems. But there are a few. And if the House would allow me to use those—and if those phones could run the Microsoft applications that all House members must use—I would do so in a second.

This is the world we live in. Monopolies, by definition, limit choice.

And often, they obliterate it.

It's ironic but also fitting that I mention Microsoft here. Although the computing company had its fair share of antitrust issues in the past, Microsoft has shown signs of syncing American priorities with its corporate planning. When Google refused to help strengthen the U.S. military by boycotting the JEDI Project, Microsoft took a strong stand for our men and women in uniform.

Not only did it participate in a program to create a visor that gives U.S. soldiers an advantage in deadly combat, but company president Brad Smith countered resistance from employees to doing business with the Department of Defense by suggesting naysayers look for other work. "We want the people of this country and especially the people who serve this country to know that we at Microsoft have their backs. They will have access to the best technology that we create."

That is the kind of default we need. One that puts our nation and its citizens first.

No matter where we each may stand on the political spectrum. I believe Americans value the free flow of information and the principles of individual liberty more than they value same-day shipping. I believe that while many Americans love social media and the convenience of search engines, the majority would be horrified by the thought of Facebook, Google, or Twitter owners deciding what subjects are permissible to discuss—or which news articles are okay to share, and which are banned. I believe that these citizens have faith in a system that gives ownership to the inventors of new ideas and businesses and doesn't let big businesses like Amazon just steal ideas or crush upstarts with impunity.

These Americans, if they remember their history lessons, know that monopolists in the late nineteenth and early twentieth centuries controlled energy, finance, steel, and commercial transportation. The difference between those monopolies and Google, Facebook, Amazon, and Apple, however, is that the new crop of corporations own the platforms that *control communication.* They control the operating systems of our phones and computers. They control the algorithms that analyze and rank search results, thereby managing who sees what content. They own and parse the personal data of all their users. They control where ads are placed and who gets to place them and for how much. These monopolies allow them to control the dissemination of information and infringe on the free flow of ideas. That means today's Big Tech monopolies threaten the core of our political system. They have the power to control the information that is available to the public and shape that information to benefit their own commercial interests and political views. When control over information in a democracy rests in the hands of only a few individuals, the results of an election can be manipulated by those individuals. With Big Tech's massive financial resources and command of critical digital media, these companies are positioned to dominate and distort not only financial and transactional marketplaces but most importantly, the marketplace of ideas. This threat to free speech is a risk that America can't afford.

In the *Politico* article, I described my decision to actively avoid using Big Tech products as a "conscience thing." Each American has the ability, perhaps the moral obligation, to act and protect our democracy from the tyranny of the Big Tech monopoly platforms. The actions may take many forms and require various levels of sacrifice. The answer can't be to ignore the problem and hope someone else takes care of the issue.

The legislation I've mentioned in this book is the heart and soul of that action. But the fight to rein in Big Tech is comprised of many actions and requires various levels of sacrifice.

We can't just passively put our fate in the hands of Big Tech because doing otherwise might be slightly more inconvenient or too inefficient or too costly—especially if the price we end up paying is the freedom of speech, the right of ownership, or keeping the entrance to the marketplaces of ideas and of commerce open. Seriously, bookmarking DuckDuckGo.com or typing it into a browser takes a second or two. Writing a congressman or -woman takes 10 minutes. Finding a retailer that has what you need at a comparable price that isn't Amazon also takes just a few minutes.

Is it less convenient?

Maybe. But fighting monopolies that crush competitors, that suppress political opinions, that ignore patent rights, that devalue and eliminate newspapers, is its own reward. And you may discover a company or shop that you love that showcases items Big Tech companies ignore.

I'm honored to speak at events in Colorado and around America. I often repeat one of my favorite quotes from John Stuart Mill. Yes, the same John Stuart Mill I mentioned in the beginning of the book who helped develop the concept of the marketplace of ideas. Mill said:

> A man who has nothing which he is willing to fight for, nothing which he cares more about than he does about his personal safety, is a miserable creature who has no chance of being free, unless made and kept so by the exertions of better men than himself. As long as justice and injustice have not terminated their ever-renewing fight for ascendancy in the affairs of mankind, human beings must be willing, when need is, to do battle for the one against the other.

For readers who feel as I do, the time has come to do more than just talk about the evils of Big Tech. It is our job to resist the Big Tech monopolies and rein in the worst offenders.

Here are a few examples of action that concerned citizens can take:

1. Seek alternatives to Big Tech products

You don't have to search using Google, shop with Amazon, socialize on Facebook. There are *many* other choices out there. Here is a starter list, beginning logically with search engine alternatives:

- **Search engines.** There are at least a dozen reputable choices: Bing.com, Webcrawler.com, and some, like Duckduckgo.com and Swisscows.com do not track your data.
- **Shopping.** There are thousands of mom-and-pop retailers on the internet. And there are exclusive online-only shops, like Etsy and UncommonGoods. There are also giant retailers, like Walmart, Target, Costco, and others with great deals on shipping and in-store pickup. Use them!
- **Video-watching.** Probably the closest competitor to YouTube is Dailymotion.com, but there are videos galore on Vimeo, Vevo (music videos), and videoshub.
- **Movie streaming services and rental stores.** Prime Video, Apple TV, and Google Play want all your entertainment dollars. But the internet is filled with terrific movie-viewing alternatives. Subscription platforms like Hulu, Disney Plus, HBOMax, Netflix, and Criterion offer all-you-can-watch deals, while à la carte movie vendors FandangoNOW and Vudu let you pick and choose individual titles. There are also plenty of free streaming movie options, such as Crackle and the amazing Internet Archive (archive.org), which offers thousands of old films, shorts, and documentaries. Finally, it's also worth mentioning that Roku, an American tech company that offers hardware and software allowing users to access and manage streaming content on their TVs, offers free movies and TV channels.
- **Music streaming.** While phones often come with Apple Music and Google Play apps installed, music streaming services abound. Spotify, of course, is the market leader,

while Tidal, Qobuz, Pandora, SiriusXM, and the all-classical Idagio also offer millions of songs for a monthly fee. There are also free sites like Soundcloud and Bandcamp that offer millions of streams for free.

- **Web hosting/cloud computing.** The cloud-based hosting and solutions business is enormous. There are small boutique companies, point-and-click platforms like Shopify, and enterprise powering companies like IBM. All offer an alternative to Amazon Web Services or Google Cloud platforms.
- **Social media.** There are many other sites to create your own online spaces. Some are text-based, like Truth Social, Parler, Reddit, and LinkedIn. Others focus on images, like Pinterest, or video clips, like Snapchat.
- **Phones.** While Google and Apple have cornered the market on smartphone operating systems, there are alternatives if you look. Meanwhile the real entrepreneurs out there should see this as a market opportunity to build something newer, less expensive, and more secure![2]

2. Change your settings!

One way to curb Big Tech's influence in our lives is to take control over the volume of data we share with these platforms. The platforms provide tools that allow for this, many of them mandated by laws such as the EU's General Data Protection Regulation and California's more recent Online Privacy Protection Act, but they do not publicize them and often the tools are buried deep in the platform's website.

Consumer Reports has a great article about changing settings on Google's web pages. You can learn how to stop location tracking, delete search history, delete web and app activity, and stop data sharing (https://www.consumerreports.org/privacy/how-to-use -google-privacy-settings/).

- **iOS tracking.** You can also stop location tracking on your iPhone. Again, *Consumer Reports* has a free helpful article on that, too. (https://www.consumerreports.org/privacy/what-the-new-iphone-tracking-setting-means-and-what-to-do-when-you-see-it).
- **Android tracking.** DuckDuckGo's android app has an option to block all app tracking on an Android device (https://www.ghacks.net/2021/12/31/my-experience-with-duckduckgos-app-tracking-protection-feature/).
- **Use VPN.** Another way to stop data tracking is to sign up for a virtual private network (VPN). These are services that connect your phone or computer to a private server that uses encryption to hide your identity—so that nobody knows who you are (unless you tell them). VPN lets users maintain more anonymity and privacy.

3. Get involved

Does your member of Congress support antitrust legislation? Find out and let them know how you feel. You may not think your voice registers, but I believe it does.

Every week, I get an Issues Summary Report from my staff. Every responsible member of the House has their staff do the same. Some weeks, my office receives thousands of emails and phone calls. My team reads each communication and listens to every call and takes notes. I study the summary carefully because it tells me what my constituents care about. I like attending events in my district and talking to voters—this Memorial Day, I attended three events and must have had conversations with 100 people. Some just wanted a picture. Some wanted to thank me for a vote or call I made. But some wanted to talk about specific issues—and that can be really impactful. One Colorado constituent who was passionate about immigration issues, shadowed me at every public event for a year. She would politely shake my hand

and talk to me for a minute or two and then move on. Over time, her dedication and insight made a real impact on me.

I share this because I want readers to know that individuals can help move mountains—and legislation. The email addresses for members of Congress may be found on each member's website. You can also go to the websites www.house.gov or www.senate .gov. The phone number for the U.S. Capitol switchboard is: (202) 224-3121. The switchboard operator can connect you to your representatives in Congress if you would like to leave a message with their staff.

4. Push the pledge!

Ask your elected leaders and favorite political organizations if they take donations from Big Tech companies. If the answer is yes, share your thoughts on the matter.

Tell them about the "Pledge for America" I took with six other members, swearing off Amazon, Apple, Facebook, Google, and Twitter campaign donations. Here are highlights of our position regarding Big Tech at the time. Every word still rings true today:

- In the public sphere, we have seen . . . the tech giants take unprecedented actions to silence political speech they disagree with, which almost exclusively means shutting down conservative speech.
- These companies are also now deserting the country that provided them the legal, commercial, and social environment to grow and thrive and are vigorously courting the communist Chinese regime in hopes of accessing cheap labor and a virtually limitless customer base.
- Whether their behavior is directed at potential competitors or at politicians and citizens they don't agree with, these companies are able to act with complete impunity because of their status as monopolies. Their respective monopolies

have given them the power to police the marketplace of ideas and dictate what information and services consumers can see and use.

5. Spread the word

Discuss with your friends, family, coworkers, and church groups. Call into radio shows. Write letters to the editor and send them to your local newspaper. Share articles. My website has three years' worth of press releases, many of which zero in on stopping Big Tech abuses. If you feel this is a significant issue—and I think most people, if they stop to think about it, value freedom—discuss monopoly power with those close to you. If they share your views, encourage them to be heard!

6. Donate

Let me repeat two very true statements: "Freedom isn't free" and "You can't run a campaign on pixie dust." I don't know any elected officials who like asking their fellow citizens for money. But this is the world we live in. Google, Apple, Facebook, and Amazon have billions of dollars at their disposal to help them consolidate power and increase market share—of everything. Consider donating your time or treasure to ensure Big Tech's monopoly power is stopped. You can donate to pro-competition candidates, think tanks, foundations, or political organizations.

7. Make your voice heard: Vote

Ultimately, politicians need to listen to voters or suffer the consequences. I urge all citizens to educate themselves on the issues and the candidates and to engage with leadership and work to get out the vote. We must use our political freedom to hold candidates and politicians accountable and ensure they uphold individual rights over the Big Tech monopolies. Ultimately, the issue of monopoly power is so critical because, at its core, any entity with the unimpeded might to influence elections, politicians, and the electorate threatens to derail democracy.

8. This is about the past and the future

Each American has the ability and, I believe, the obligation to take action and protect our democracy from the tyranny of the Big Tech monopoly platforms. As I've documented, the United States of America was formed in large part to counteract a vicious, oppressive monopoly. We must carry that lesson as we go forward. Combatting the current digital cartels will take many forms and require various levels of sacrifice. The answer can't be to ignore the problem and hope someone else takes care of the issue. That is why I have spent the last three years of my life focusing on these issues.

It is easy to conclude that individual consumers do not influence the market very much. But I believe the opposite is true. The market gets its power from many individuals acting in unison. We are not alone. Both the Senate and the House—two ideologically war-torn bodies—have united in a congressional miracle to try and curb Big Tech. So, we are not alone in recognizing the threat digital monopolies present to our democracy. Don't buy products from immoral companies that discriminate based on political viewpoints, silence newspapers, and stymie competition. Encourage your employer, friends, and neighbors to avoid the Big Tech monopolists. You can live without next-day delivery, nano-second search results, and endless hours on a social media platform.

Many Americans don't have access to a book like this or other information about Big Tech. Share information with the people around you—at work, at church, online, at the supermarket. And remember: before social media, before the internet, before cable

TV, or radio or newspapers, the oldest and most trusted form of news distribution was something called word-of-mouth. Now more than ever, let it roar.

Epilogue

In this book we have traveled together from the days of the Boston Tea Party to the robber barons of the mid-nineteenth century and arrived at the greatest domestic threat to our democracy—Big Tech monopoly power over the dissemination of information. History has shown us how a few courageous citizens can defeat the greatest military power on earth or defeat the most powerful business leaders in America's industrial revolution. Now history—and our current chilling reality—must inspire us to protect our democracy.

There is, of course, a legislative response that we must fight for. And I want to update you on that. One year after our 29-hour marathon antitrust hearing in which we passed six new laws out of committee, we have very little to show for it. The Democratic-controlled Congress has produced nothing. In fact, only one of the six bipartisan bills to rein in the monopoly power of Google, Amazon, Facebook, and Apple—stood a chance of getting a vote in both the House and Senate by the end of 2022.

One bill.

We spent 16 months investigating. We had multiple hearings. We combed through thousands of emails. We crossed the aisle over and over to find common ground and issue a bipartisan bill. We sweated out a 29-hour hearing. We did great work trying to return stability and balance to America. To protect free speech, free competition, and open access to the marketplace of ideas. To ensure innovation—one of the hallmarks of America—is never stifled.

Somehow, with Democratic control of the Oval Office, the Senate and the House, all that work will amount, ideally, to the passage of a single law.

Let me repeat that:

One bill.

Outrageous.

As I write this, there are stories about Big Tech's all-out war against our proposed legislation. The monopoly-loving corporations reportedly launched a $36 million ad campaign to criticize our six bills. Advocacy groups supporting our antitrust measures spent $200,000.

This again speaks to the imbalance of power that should be a call to action.

Monopoly power over the dissemination of information begets more power and more censorship. And that appears to be what has happened here.

Regardless of ad budgets and influence campaigns, the fact is two people controlled the bills that members of Congress voted on in 2021 and 2022: the House speaker and the Senate leader. Their names are Nancy Pelosi and Chuck Schumer. As I documented in Chapter 11, both leaders have strong family and business ties—direct and indirect—to Big Tech. The decision not to allow Congress to address and correct these vast monopolies lies with them and, to a lesser degree, with President Biden, who could have applied pressure to his party compatriot leaders.

Even if Pelosi and Schumer allowed a vote on The American Innovation and Choice Online Act—the bill they say they were considering—and if it passed, it would not have been enough. An enormous, nation-strengthening and nation-saving opportunity was lost.

I do not mean to denigrate the Innovation and Choice Online Act; it is a critical bill. As I've reported, it would prevent Big Tech from self-preferencing their products, services, or lines of business. It would also bar anticompetitive and discriminatory platform regulations, allow preinstalled apps to be removed and default settings changed, ban retaliation, outlaw abuse and misuse of nonpublic data for a competitive advantage, and more. That is huge.

But I want to be very clear. These monopolies are massive. Undoing them would take more than one piece of legislation. Six concepts need to be addressed in Congress to combat Big Tech monopolies and the vast anticompetitive challenges they present to e-commerce, free speech, and innovation.

1. **Clarify rules against predatory pricing by monopolies.** Lower prices are great for consumers in the short term. But the long-term implications are frightening for all when pricing becomes a tool to eliminate competition and stifle innovation. When that happens, society at large stuffers. As we saw with Amazon's diaper war with Quidsi, Amazon used pricing to wound and eliminate a true e-commerce innovator. Now, due to the vast profits reaped by its web services company, Amazon has the bankroll to target almost any competitor it chooses and put it out of business. But Jeff Bezos's empire isn't the only looming predator. All four Big Tech firms have the cash to diversify into new markets and sustain huge losses if that fits their strategic goal of maintaining monopoly power.

2. **Require Big Tech monopolies to prove mergers and acquisitions are pro-competitive.** Washington's own

watchdogs have admitted failure in policing mergers. Congress must help them and provide funding for closer scrutiny and direction for more rigorous tests. Does a change in ownership give a company too much power? Will an acquisition drive competition and spur further innovation or eliminate it, shrinking the marketplace of ideas and investment with it?

3. **Reform the Patent Trial and Appeal Board.** The war against intellectual property and innovation began in earnest when the Obama administration embraced serving Big Tech's whims as a matter of policy. The PTAB has turned hundreds of years of precedent protecting the rights of inventors on its head. Why? One reason was the rise of patent trolls—bad actors who obtain patents and weaponize them with often frivolous but costly suits. These bad actors need to be dealt with. But undoing legitimate patents as a matter of unofficial policy so that Big Tech can grab and profit from the work of others is even worse. This is what PTAB does. It discourages American innovation and allows economic and political rivals, like China, to use groundbreaking American work without compensating its creators. It must be stopped. And patent protection must be ensured to incentivize American innovation.

4. **Waive antitrust laws so newspapers can negotiate with Google.** Newspapers—and news-gathering organizations—are extremely valuable to society. They are also precious to Google, providing content for the search engine to display. Search results pages displaying links to news articles create multiple wins for Google: advertising opportunities on the search results pages that link to articles and user-interest data created with each subsequent click. Newspapers need the right to band together and

seek fair compensation for their content—content that has allowed Google to become the most profitable and influential disseminator of information on the planet. In this instance, we must allow corporate collective bargaining to counteract an industry-killing behemoth.

5. **Prohibit Big Tech companies from controlling buy, sell, and auctions with ads.** Google isn't just the biggest distributor of news on the planet; it's also the biggest distributor of digital advertising. The case for an antitrust action is pretty clear: numerous suits charge Google runs the world's largest ad exchange, charging ad buyers commission fees while also representing and charging ad space sellers, and administering a third exchange usage fee. If allegations that Google entered into a deal with Facebook to prevent it from using a rival header bidding ad platform and gave it favorable terms prove true, well how much more evidence of anticompetitive behavior do we need? Furthermore, this isn't just about controlling a market from a business standpoint. Control the sale and display of ads and you are controlling the display of speech. Google isn't too big to fail. Its behavior suggests it is so big and controls all sides of the market that it makes everyone else fail.

6. **Prohibit the self-preferencing from hindering free speech.** I'm a big fan of the American Innovation and Choice Online Act. But although it tackles self-preferencing, it doesn't go far enough. Self-preferencing isn't just about favoring one product or service over another for economic gain or increased market share. It's also about favoring one idea, candidate, or news story over another because doing so serves the needs of a company or its owners or allows that company to curry favor with a political group. This is an enormous threat. Amazon

suppressing a book or a movie, Google burying links that offer a different perspective to a news event, Facebook deleting or flagging a post, Apple refusing to offer the Bible in its app store—these are examples of *preferring* censorship to open society, of limiting the marketplace of ideas. At this point, I know it's no surprise when I say nothing is more important to the success of America than promoting that marketplace and ensuring all have access. It is Congress's job to see that through.

It is impossible to predict the outcome of elections—as Trump proved on November 9, 2016. He shocked Hillary Clinton and the entire D.C. establishment, and ascended to the Oval Office by reshaping the Republican Party, shaking off the label of caring only about the wealthy and big corporations. He showed Republicans that we can win by representing the blue-collar worker, by fighting against outsourcing jobs to China, by bringing the supply chain home.

The 2022 midterm elections returned the Republican Party to power on Capitol Hill. How this will impact efforts to restrict monopoly power is anyone's guess.

Will the new Republican majority in the House continue with the Trump vision of protecting our civil rights and of looking out for all citizens of our nation?

I know which side of the question I plan to represent. Protecting the right to speak freely—the marketplace of ideas— and upholding our foundational antimonopoly legacy is a never-ending story. This book has chronicled a very real threat to our democracy, and I hope that you join me, the other members of Congress, and conservative leaders nationwide who are promoting competition, advancing free speech, and fighting the tyranny of Big Tech.

Notes

Chapter 1

1. For those wondering what I did to earn Olbermann's ire, he anointed me for his award because I had asked aloud if the private sector might run a Veterans Administration hospital more efficiently than the public sector. This simple question incensed him. I will note, however, that a few years after Olbermann tried to call me out, there were widespread reports of veterans languishing on waiting lists at VA hospitals and dying before they could see a doctor. I was right to ask the questions I did about optimizing and improving VA health-care management.

2. Some of you may wonder why I chose to cite John Stuart Mill for the proposition that speech should suffer from as little regulation as possible. While Mill's social and economic policies would make Karl Marx blush, he was right about speech. Today our Marxist/progressive/liberal friends don't want to engage in open debate. They focus on winning the battle of ideas by censoring all who dare to disagree with them. Not so in Mill's time. He encouraged debate, even if he was on the wrong side of economic history.

3. Oliver Wendell Holmes, https://firstamendmentwatch.org/history-speaks-holmes-dissenting-abrams-v-united-states-1919/.

4. https://www.opensecrets.org/federal-lobbying/top-spenders.

5. https://nypost.com/2021/11/28/media-helped-hide-the-real-joe-biden-by-censoring-hunter-stories-devine/.

6. https://pjmedia.com/news-and-politics/paula-bolyard/2018/08/25/google-search-results-show-pervasive-anti-trump-anti-conservative-bias-n60450.

7. https://ballotpedia.org/Elected_officials_suspended_or_banned_from_social_media_platforms.

Chapter 2

1. For 170 years or so, the initial 70,000-pound investment to start the East India Company—like buying shares of Amazon stock at $2 and change in 1998—paid off with mind-boggling returns.
2. https://thehill.com/opinion/technology/3469063-competition-is-not-a-click-away-the-consumer-welfare-standard-is-failing-consumers/.
3. John Dickinson, *Two Letters on the Tea Tax*, Carlisle, MA: Applewood Books, 2009, pp. 459–460.
4. https://founders.archives.gov/documents/Jefferson/01-12-02-0454.
5. https://scholarlycommons.law.northwestern.edu/cgi/viewcontent.cgi?article=1213&context=facultyworkingpapers, p. 35.
6. William T. Hutchinson, et al. (eds.), *The Papers of James Madison*, Chicago: University of Chicago Press, 1962–1977.
7. Steven G. Calabresi and Larissa Price, "Monopolies and the Constitution: A History of Crony Capitalism" (2012), Faculty Working Papers, Paper 214, https://scholarlycommons.law.northwestern.edu/cgi/viewcontent.cgi?article=1213&context=facultyworkingpapers.
8. https://www.heritage.org/constitution/#!/articles/4/essays/122/privileges-and-immunities-clause.
9. A clarification is in order: while Franklin wrote these words—and printed them in his brother's newspaper—he attributed them to Silence Dogood, an elderly widow. You could look it up at: https://www.thefire.org/for-the-fourth-benjamin-franklin-kind-of-on-freedom-of-speech/.
10. https://www.wsj.com/articles/the-great-firewall-of-china-review-the-chinese-cyber-padlock-11552245606.
11. https://www.christianitydaily.com/articles/13391/20210924/big-tech-censors-conservatives-in-congress-more-than-it-does-liberals-53-1.htm.
12. Menahem Blondheim, "Rehearsal for Media Regulation: Congress Versus the TelegraphNews Monopoly," 1866–1900, *Federal Communications Law Journal*, vol 56, issue 2, March, 2004.

Chapter 3

1. https://www.motherjones.com/politics/2016/10/veteran-spy-gave-fbi-info-alleging-russian-operation-cultivate-donald-trump/.
2. https://www.nytimes.com/2016/11/01/us/politics/fbi-russia-election-donald-trump.html?_r=0.
3. https://www.buzzfeednews.com/article/kenbensinger/these-reports-allege-trump-has-deep-ties-to-russia.

4. https://www.justice.gov/storage/120919-examination.pdf, p. 187.

5. https://www.washingtonpost.com/politics/2021/10/18/big-hole
    -christopher-steeles-defense-himself/.

6. https://www.npr.org/2019/02/28/699118969/rep-eric-swalwell-there
    -is-at-least-one-indictment-waiting-for-president-trump.

7. Adam Smith, *An Enquiry into the Nature and Causes of the Wealth
    of Nations*, London: Strahan & Cadell, 1776, p. 15.

8. Friedrich August von Hayek, *The Fatal Conceit: The Errors of
    Socialism* (The Collected Works of F. A. Hayek, Book 1), William
    Warren Bartley (ed.), Chicago: University of Chicago Press, 1988,
    p. 14.

9. https://news.harvard.edu/gazette/story/2019/03/harvard-professor
    -says-surveillance-capitalism-is-undermining-democracy/.

10. https://www.technologyreview.com/2012/06/13/185690/what
    -facebook-knows/.

Chapter 4

1. https://time.com/5687134/trump-universal-postal-union-deal/.

2. https://morningchalkup.com/2018/07/30/rogue-fitness-is-getting
    -sued-for-patent-infringement/.

3. https://www.uspto.gov/patents/laws/leahy-smith-america-invents
    -act-implementation.

4. https://www.heritage.org/courts/commentary/the-supreme-court
    -must-protect-patent-rights.

5. Janice M. Mueller, *Aspen Treatise for Patent Law*, Boston: Aspen
    Publishing, 2020.

6. https://www.ipwatchdog.com/2018/03/21/how-google-and-big-tech
    -killed-the-u-s-patent-system/id=95080/.

7. https://www.politico.com/story/2012/05/tech-world-cools-to-obama
    -076651.

8. https://casetext.com/admin-law/jump-rope-systems-llc.

9. https://www.ipwatchdog.com/2018/03/21/how-google-and-big-tech
    -killed-the-u-s-patent-system/id=95080/.

10. https://patentlyo.com/patent/2015/06/america-invents-trillion.html.

11. https://www.pewresearch.org/internet/2009/04/15/the-internets
    -role-in-campaign-2008/.

12. https://web.archive.org/web/20210109031942/https://www
    .thetrumparchive.com/.

13. https://www.cnn.com/2020/12/18/politics/trump-presidency-by-the
    -numbers/index.html.

14. https://www.opensecrets.org/online-ads.

Chapter 5

1. China, no matter what the CCP says, does not have a free and open economy. The government monitors the disbursement of all funds going overseas. It has a seat on the board of every corporation in the country. It can and has replaced CEOs of major companies, including Jack Ma, the founder of Alibaba. It understands that controlling finance and wealth allows it to control behavior. This idea is at the root of both its domestic and foreign policies.
2. http://www.newsmediaalliance.org/wp-content/uploads/2019/06/Google-Benefit-from-News-Content-SUMMARY.pdf.
3. https://searchengineland.com/google-q4-2021-earnings-379735.
4. https://www.pewresearch.org/fact-tank/2021/07/13/u-s-newsroom-employment-has-fallen-26-since-2008/.
5. https://www.usnewsdeserts.com/reports/expanding-news-desert/loss-of-local-news/loss-newspapers-readers/.
6. https://www.poynter.org/locally/2021/the-coronavirus-has-closed-more-than-100-local-newsrooms-across-america-and-counting/.
7. Irina Ivanova, "Newsrooms Look to Reclaim Ad Dollars from Big Tech," CBS News, June 11, 2019, https://www.cbsnews.com/news/news-media-alliance-blames-big-tech-for-taking-their-ad-dollars/.
8. https://mattersoffact.org/blog/googles-ad-monopoly.
9. https://www.nytimes.com/2020/09/21/technology/google-doubleclick-antitrust-ads.html.
10. https://www.justice.gov/opa/pr/justice-department-sues-monopolist-google-violating-antitrust-laws.
11. Ibid.
12. https://www.theatlantic.com/politics/archive/2021/10/gannett-local-newspaper-hawk-eye-iowa/619847/.
13. https://www.cnbc.com/2017/06/27/eu-hits-google-with-a-record-antitrust-fine-of-2-point-7-billion.html.

Chapter 6

1. https://www.youtube.com/watch?v=qOMeGBy-t7g.
2. Dieter Bohn, "The Apple App Store: A Brief History of Major Policy Changes," *The Verge*, September 10, 2021, https://www.theverge.com/22667242/apple-app-store-major-policy-changes-history.
3. "A Timeline: How We Got Here," Timetoplayfair.com, https://www.timetoplayfair.com/timeline/.
4. Chris Welch, "Spotify Urges iPhone Customers to Stop Paying Through Apple's App Store," *The Verge*, July 8, 2015, https://www.theverge.com/2015/7/8/8913105/spotify-apple-app-store-email.

5. "A Timeline: How We Got Here," Timetoplayfair.com, https://
    www.timetoplayfair.com/timeline/.
6. Russell Brandom, "Apple Must Allow Other Forms of In-App
    Purchase, Rules Judge in Epic *v*. Apple," *The Verge*, September 10,
    2021, https://www.theverge.com/2021/9/10/22662320/epic-apple
    -ruling-injunction-judge-court-app-store.
7. https://developer.apple.com/app-store/review/guidelines/#scraping
    -aggregation.
8. https://files.constantcontact.com/60ec52f3801/7ea6b116-bf93-452f
    -a48d-acc86c481c3c.pdf.

Chapter 7

1. https://www.vanityfair.com/news/2019/04/inside-the-mark
    -zuckerberg-winklevoss-twins-cage-match.
2. https://officechai.com/stories/incredible-facebook-emails-show
    -mark-zuckerberg-approached-instagram-acquisition-2012/.
3. https://judiciary.house.gov/uploadedfiles/0006331600063321.pdf.
4. https://www.ftc.gov/sites/default/files/documents/closing_letters/
    facebook-inc./instagram-inc./120822barnettfacebookcltr.pdf.
5. It is worth noting that WhatsApp gained its massive user base by
    vowing to protect user privacy and data from ad targeting. And that
    Facebook reportedly decided to buy WhatsApp after discovering
    users sent 8.2 billion messages a day via the rival app compared
    to 3.5 million on Facebook's mobile Messenger. https://epic.org/
    documents/in-re-whatsapp/.
6. Georgia Wells, Jeff Horwitz, Deepa Seetharaman, "Facebook Knows
    Instagram Is Toxic for Teen Girls, Company Documents Show: Its
    Own In-Depth Research Shows a Significant Teen Mental-Health
    Issue That Facebook Plays Down in Public," *Wall Street Journal*
    (online), September 14, 2021, https://www.wsj.com/articles/facebook
    -knows-instagram-is-toxic-for-teen-girls-company-documents
    -show-11631620739.
7. https://www.npr.org/2021/05/18/990234501/facebook-calls-links
    -to-depression-inconclusive-these-researchers-disagree.
8. https://www.washingtonpost.com/technology/interactive/2021/
    amazon-apple-facebook-google-acquisitions/.
9. https://nypost.com/2021/07/15/white-house-flagging-posts-for
    -facebook-to-censor-due-to-covid-19-misinformation/.
10. https://www.facebook.com/help/230764881494641/.
11. https://www.washingtonexaminer.com/opinion/op-eds/the-cdc-has
    -lost-all-credibility.

Chapter 8

1. https://www.iab.com/news/digital-advertising-soared-35-to-189
-billion-in-2021-according-to-the-iab-internet-advertising-revenue
-report/.
2. https://variety.com/vip/2021-tv-ad-spend-trends-movie-marketing
-rebounds-networks-decline-1235143104/.
3. https://www.ibisworld.com/us/bed/print-advertising
-expenditure/4676/.
4. https://www.theverge.com/2021/2/8/22272230/west-virginia
-newspaper-google-facebook-lawsuit-digital-ad-revenue.
5. https://www.editorandpublisher.com/stories/hd-media-files-first-of
-its-kind-antitrust-lawsuit-against-google-and-facebook,185360#.
6. https://www.scribd.com/document/492607988/Complaint-HD
-Media-Co-LLC-v-Google.
7. https://www.texasattorneygeneral.gov/sites/default/files/images/
admin/2020/Press/20201216_1%20Complaint%20(Redacted).pdf.
8. https://twitter.com/TXAG/status/1339283520099856384.
9. Public service reminder: In Chapter 5 I noted that at least one of
the FCC voting members, William Kovacic, who approved Google's
DoubleClick acquisition, now admits he made a mistake. The deal
should never have been approved.
10. Tracy Ryan and Jeff Horwitz, "Inside the Google-Facebook Ad Deal
at the Heart of a Price-Fixing Lawsuit," *Wall Street Journal* (online),
December 29, 2020, https://www.wsj.com/articles/inside-the-google
-facebook-ad-deal-at-the-heart-of-a-price-fixing-lawsuit
-11609254758.
11. https://storage.courtlistener.com/recap/gov.uscourts.nysd.564903/
gov.uscourts.nysd.564903.152.0.pdf, p. 6.
12. Sam Schechner, Kirsten Grind, and John West, "Searching for Video?
Google Pushes YouTube Over Rivals," *Wall Street Journal* (online),
July 14, 2020, https://www.wsj.com/articles/google-steers-users-to
-youtube-over-rivals-11594745232.
13. https://www.reuters.com/article/us-google-privacy-france/france
-fines-google-57-million-for-european-privacy-rule-breach
-idUSKCN1PF208.
14. https://themarkup.org/ask-the-markup/2021/09/02/what-does-it
-actually-mean-when-a-company-says-we-do-not-sell-your-data#.
15. https://insights.hgpresearch.com/who-owns-personal-data-gdpr
-vs-usa.

Chapter 9

1. https://judiciary.house.gov/uploadedfiles/00151722.pdf.
2. https://slate.com/technology/2013/10/amazon-book-how-jeff-bezos
    -went-thermonuclear-on-diapers-com.html.
3. Timothy B. Lee, "Emails Detail Amazon's Plan to Crush a Startup
    Rival with Price Cuts," *Ars Technica*, July 7, 2020, https://arstechnica
    .com/tech-policy/2020/07/emails-detail-amazons-plan-to-crush-a
    -startup-rival-with-price-cuts/.
4. https://www.wsj.com/articles/split-amazon-in-two-prime-web
    -services-aws-logistics-third-party-earnings-report-consumers
    -antitrust-11644249482.
5. Dana Mattioli, "Amazon Scooped up Data from Its Own Sellers to
    Launch Competing Products," *Wall Street Journal* (online), April 23,
    2020, https://www.wsj.com/articles/amazon-scooped-up-data-from
    -its-own-sellers-to-launch-competing-products-11587650015.
6. https://www.reuters.com/investigates/special-report/amazon-india
    -rigging/.
7. Jason Riley, "Will Amazon Suppress the True Michael Brown Story?,"
    Wall Street Journal, New York, October 15, 2020, A16.
8. Caitlin O'Kane, "Amazon Has Stopped Selling Books That Frame
    LGBTQ+ Identities as Mental Illnesses," CBS News, March 12, 2021.
9. Committee of the Judiciary, "Amazon Referral Letter," March 9, 2022,
    https://judiciary.house.gov/uploadedfiles/hjc_referral_--_amazon.pdf,
    p. 3.
10. Just look at the command economies of China, North Korea, and
    Russia. These are among the most sinister surveillance states on the
    planet; they are not free-market economies. In China, the Communist
    Party controls the flow of cash leaving the country, a Communist
    Party member sits on the board of every single Chinese firm, and
    there is a law on the books stating that the CCP owns all data created
    within its borders—every record of every sales transaction, every text,
    every post. Sure, there are stock markets in China, but the totalitarian
    power of the CCP is the ultimate authority, not the market.

Chapter 10

1. https://www.rev.com/blog/transcripts/big-tech-antitrust-hearing
    -full-transcript-july-29.
2. https://www.nytimes.com/2018/12/16/world/asia/xinjiang-china
    -forced-labor-camps-uighurs.html.
3. https://www.nytimes.com/2019/10/09/technology/apple-hong-kong
    -app.html.

4. https://s2.q4cdn.com/470004039/files/doc_downloads/gov_docs/
2021/03/Our-Commitment-to-Human-Rights_Final-copy
-(updated-links-Feb-2021).pdf.

5. https://www.businessinsider.com/apple-takes-down-quran-bible
-jehovahs-witenss-apps-in-china-2021-10.

6. https://thediplomat.com/2017/06/chinas-cybersecurity-law-what
-you-need-to-know/.

7. https://cloud.google.com/blog/topics/inside-google-cloud/update
-on-google-clouds-work-with-the-us-government.

8. https://about.google/human-rights/.

9. https://www.humanrights.com/what-are-human-rights/universal
-declaration-of-human-rights/articles-11-20.html.

10. Robert Spalding with Seth Kaufman, *Stealth War*, Portfolio Books,
2019, pp. 155–159.

Chapter 11

1. "Top Spenders," OpenSecrets.org, https://www.opensecrets.org/
federal-lobbying/top-spenders?

2. Lydia Moynihan, "Major Antitrust Adversary in Congress Has
Daughter on Google's Legal Team," *New York Post*, December 7,
2021.

3. Celine Castronuovo, "Pelosi Husband Won Big on Alphabet Stock,"
The Hill, July 8, 2021.

4. Lydia Moynihan. "Schumer's Duaghters Work for Amazon,
Facebook" *New York Post*, January 18, 2022, https://nypost.com/
2022/01/18/schumers-daughters-work-for-amazon-facebook-as-he
-holds-power-over-antitrust-bill/.

5. "Sen. Cruz: Big Tech Poses the Single Greatest Threat to Our Free
Speech & Democracy," Cruz.senate.gov, October, 1 2020. https://
www.cruz.senate.gov/newsroom/press-releases/sen-cruz-big-tech
-poses-the-single-greatest-threat-to-our-free-speech-and-democracy.

6. Ellen Terrell, "When a Quote Is Not (exactly) a Quote: The Business
of America Is Business Edition," Library of Congress, January 17,
2019.

7. https://www.presidency.ucsb.edu/documents/address-the-american
-society-newspaper-editors-washington-dc.

Chapter 12

1. Nancy Scola, "The Congressman Who Doesn't Use Google," *Politico*,
December 20, 2021, https://www.politico.com/news/magazine/
2021/12/20/ken-buck-congressman-google-525282.

2. Case in point: In 2022 I met with Andy Yen, CEO and founder of the encrypted email service ProtonMail. I learned that his privacy-focused start-up threatened Google so much that at one point the company vanished from Google searches. He also told me that when ProtonMail users emailed Gmail accounts, their communications were automatically funneled into Gmail spam folders. The company had to threaten legal action to get Gmail to start doing the right thing.

Acknowledgments

I want to thank the dozens of courageous colleagues, small business owners, investors, tech executives, scholars, and hardworking Americans who took the time to share their stories with me about how they were victimized, censored, canceled, and silenced by Big Tech. You inspired me and many other Americans to stand up to the four most powerful corporations in the history of the world so that we can preserve the freedom to express ourselves without being muzzled by the social media overlords.

I am also grateful for the guidance, advice, and patience of Shonda Werry during the arduous process of writing and rewriting each paragraph of this book. Ms. Werry applied her broad background in conservative activism to constantly challenge the proper balance for government enforcement of antitrust laws to increase competition without criticizing success or protecting inefficient or ineffective businesses.

While not directly involved in writing this book, I want to recognize the tremendous work of my friend and chief of staff, James Braid, along with Zach Mendelovici, Rachel Bissex, and Slade Bond for their tireless work on behalf of the American public. They are overworked, underpaid, and too often ignored despite being on the frontlines in the battle with Big Tech.

I want to thank Chris Ruddy, CEO of Newsmax, for his suggestion to write this book. Many thanks to Mary Glenn, publisher at Humanix Books, and her great team for making this book possible in a condensed time frame.

Finally, to Cody and Kaitlin, you are the joys of my life and I love you beyond words on a page.

ABOUT THE AUTHOR

KEN BUCK graduated from Princeton University in 1981 and the University of Wyoming Law School in 1985. Ken worked his way through school as a janitor, truck driver, furniture mover, ranch hand, and high school football coach.

After law school, Ken Buck worked for Congressman Dick Cheney (R-WY) on the Iran-Contra investigation and then became a prosecutor with the U.S. Department of Justice. In 1990, Ken joined the Colorado U.S. Attorney's Office where he became the Chief of the Criminal Division.

Ken Buck was elected district attorney three times in Weld County, Colorado (2004, 2008, 2012). In 2014, Ken was elected to the U.S. House of Representatives for the Fourth Congressional District in Colorado. He is currently a member of the Judiciary Committee and serves as the ranking Republican on the Antitrust, Commercial, and Administrative Law Subcommittee. Ken is also on the Foreign Affairs Committee.

He is the proud parent of Cody and Kaitlin and grandparent to Bear, Sugar Ray, Dubya, T-Man, Collins, and Doe.